RSC Pap

KU-346-762

FOOD FLAVOURS
Biology and Chemistry

CAROLYN FISHER* and THOMAS R. SCOTT

*Departments of Animal and Food Sciences and Psychology
University of Delaware
Newark DE 19716, USA*

**Present address: McCormick and Company Inc.,
202 Wight Avenue, Hunt Valley, MD 21031, USA*

THE ROYAL
SOCIETY OF
CHEMISTRY
Information
Services

664.5 AS

BRN: 192469

AC		KE	X
AT		ME	
CL		WM	
FH			
HI			

ISBN 0-85404-538-4
A catalogue record for this book is available from the British Library

© The Royal Society of Chemistry 1997

All rights reserved.

Apart from any fair dealing for the purposes of research or private study, or criticism or review as permitted under the terms of the UK Copyright, Designs and Patents Act, 1988, this publication may not be reproduced, stored or transmitted, in any form or by any means, without the prior permission in writing of The Royal Society of Chemistry, or in the case of reprographic reproduction only in accordance with the terms of the licences issued by the Copyright Licensing Agency in the UK, or in accordance with the terms of the licences issued by the appropriate Reproduction Rights Organization outside the UK. Enquiries concerning reproduction outside the terms stated here should be sent to The Royal Society of Chemistry at the address printed on this page.

Published by The Royal Society of Chemistry, Thomas Graham House, Science Park, Milton Road, Cambridge CB4 4WF, UK

Typeset by Vision Typesetting Manchester
Printed by Athenaeum Press Ltd, Gateshead, Tyne and Wear, UK

Preface

This book is designed for students of food flavours, and for those in related fields who seek an integrated overview of this area. Most books on flavours are edited contributions of individual scientists, each addressing a specialized aspect of the discipline. Here, we offer the joint perspectives of a flavour biochemist (CLF) and a neuroscientist (TRS), permitting a full sweep from synthetic chemistry through the chemical senses to the hedonic reactions of human tasters.

The purpose is to offer a deeper understanding of the chemical products generated in the food industry, and the biological and psychological reactions to them. By this means, our goal is to help position others to achieve a fuller understanding of current flavours and to discover new compounds to serve the flavour industry.

This book is presented in five chapters. The first offers an introduction to the complexities flavour scientists encounter. The second provides the chemical background on flavour compounds. In the third chapter, we describe the anatomy and physiology of the chemical sensory systems: olfaction and taste. In the fourth, we detail the sensory and the instrumental analyses of flavourants. Finally, we offer five complex flavour problems, and the methods by which students may be directed toward their solutions.

Contents

Acknowledgements

We would like to thank our spouses and families for their encouragement during the writing. Special mention should be given to Dr Fisher's son, Dan Fisher, who helped by scanning the figures into the computer and manipulating them.

We would like to acknowledge the help of those who gave their time to read the manuscript and provide very valuable comments: Dr Chi-Tang Ho, Dr Herbert Stone, Dr Ann C. Noble, Mr Paul Todd, and Ms Leora Hatchwell. Special mention must be given to Dr Chi-Tang Ho, who so graciously allowed Dr Fisher to peruse his flavour chemistry lecture notes during the formative stages of this book.

Dr Fisher would also like thank her students of the spring semesters of 1995 and 1996 for reading the manuscript drafts as they evolved and working through the problems and/or case studies that are included in Chapter 5.

We dedicate this book to Leonard Phillip Fisher and Bonnie Kime Scott.

Chapter 1

Introduction – Problems in Flavour Research

The inspiration for this chapter comes from the first chapter in *Flavour Research: Recent Advances*, by Flath *et al.*,[1] which has the same name.

DEFINITION OF FLAVOUR

What is flavour? There are two main definitions of flavour which depend upon the viewpoint of the definer. Flavour can refer to a biological perception, such that it is the sensation produced by a material taken in the mouth, or flavour can refer to an attribute of the material being perceived. The attribute is the aggregate of the characteristics of the material that produces the sensation of flavour. Flavour is perceived principally by the aroma receptors in the nose and taste receptors in the mouth. However, flavour descriptors, such as hot, pungent and biting, are also given to sensations received by the general pain, tactile, and temperature receptors in the mouth, nose and eyes.

Whether flavour refers to the chemicals responsible for the stimulation or the biological receptor stimulation itself, is immaterial to the consumer of foods. Consumers consider flavour one of the three main sensory properties decisive in their selection, acceptance, and ingestion of a particular food. The other two sensory properties are appearance and texture (Figure 1.1). We are all familiar with the basic five senses: sight (eyes), taste (tongue), odour (nose), hearing (ears) and touch (fingers, mouthfeel). The sense of touch, giving mouthfeel, can be broken down into three sensations: pressure, trigeminal and kinaesthesis. Pressure represents the feeling when force is applied over the surface of the food, trigeminal refers to a pain sensation and kinaesthesis denotes feedback from masticatory muscles during chewing.

1

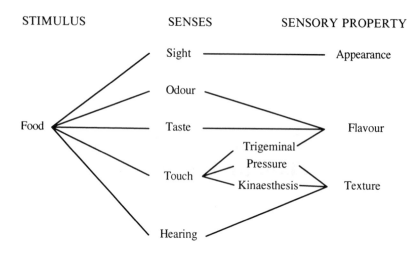

Figure 1.1   *Relationship of the five senses with sensory properties.*

CLASSIFICATION OF FOOD FLAVOURS

Flavours can be classified by the general sensations that one feels when eating different foods (Figure 1.2). As observed in Figure 1.1, flavour comes from three different sensations: taste, trigeminal and aroma (odour). It is generally agreed that taste sensations are divided into four major categories: saltiness, sweetness, sourness and bitterness. However, some Japanese scientists also include a fifth category called *umami* (savoury) that can be represented by the flavour of glutamate. Trigeminal sensations give us the descriptors of astringency, pungency and cooling. Both taste and trigeminal sensations occur upon contact with food in the mouth, as most substances which produce these flavours are non-volatile, polar, and water-soluble. For aroma sensations to occur, an aromatic compound must be sufficiently volatile to allow detection at a distance. The physical interaction between the volatile compound and the receptor site occurs in the nasal passages. Those molecules that reach the olfactory receptors, either via the nasal passage or oral passageway, trigger the odorous sensations.

However, food flavourants are usually classified by the food sources in which they are normally detected (Table 1.1) because more than one flavour sensation is usually triggered by a food flavourant. Given a specific flavourant, the food industry wants to know what type of image the average consumer will envision when he or she encounters it. For example, celery flavourant (from an extract of celery seed) used in a soup is bitter with a floral

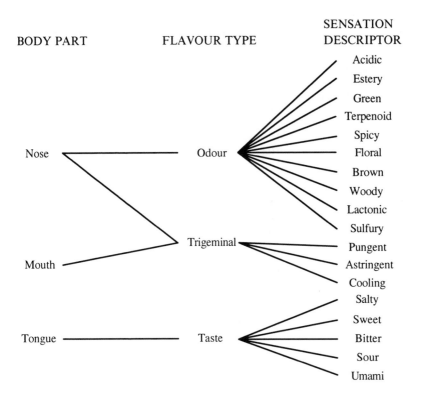

Figure 1.2 *Correlation of flavour types with sensation descriptors.*

aroma, but to an average consumer this flavourant just elicits the thought of celery soup.

The problem with using food sources to classify flavours is that flavours may vary with the history of the food source. For example, fresh cabbage has a quite different aroma than cooked cabbage and sauerkraut is a vastly different olfactory and gustatory experience! Thus classifying flavours by food source is somewhat arbitrary, with the processing method frequently denoted in the descriptive name of the flavour.

Fruit Flavours

The *tastes* of fruit are a blend of the sweetness due to sugars (such as glucose, fructose and sucrose) and the sourness of organic acids (such as citric and malic). However, it is the *aromas* of the different volatile components of fruits that allow us to distinguish among them. When one's sense of smell is

Table 1.1 *Classification of food flavours*

Flavour Class	Subdivision	Example
Fruit flavours	Citrus (terpene)	Grapefruit, orange
	Non-citrus (non-terpene)	Apple, raspberry, banana
Vegetable flavours	Fresh	Lettuce, celery
	Dried	Tomato leather, tobacco
Spice flavours	Aromatic	Cinnamon, peppermint
	Lachrymatory	Onion, garlic
	Hot	Pepper, ginger
Beverage flavours	Unfermented	Juices, milk
	Fermented	Wine, beer, tea
	Compounded	Soft drinks, cordials
Meat flavours	Mammal	Beef, lamb, pork
	Fish	Salmon, menhaden
	Fowl	Chicken, turkey
Fat flavours	Vegetable	Olive oil, soybean oil
	Animal	Lard, tallow, butter
Cooked flavours	Broth	Beef bouillon
	Vegetable	Peas, potatoes, beans
	Fruit	Marmalade, jelly
Empyreumatic flavours	Smoky	Ham, kippers
	Broiled, fried	Processed meats
	Roasted, toasted, baked	Coffee, snack foods, breakfast cereals, bread
Stench flavours	Fermented	Blue cheese
	Oxidized	Spoiled fish

eliminated (temporarily having a stuffy nose from a cold), it is extremely difficult to distinguish between onions and apples.

A typical fruit may have well over a hundred different volatile components, but in total, these compose only a few parts per million of the entire fruit.

Fruit aromas vary widely. Citrus, such as grapefruit, orange, lemon and lime, are rich in terpenoids whereas most non-citrus fruits, such as apple, raspberry, cranberry and banana, are characterized by esters and aldehydes.

Vegetable Flavours

Most cultivated vegetables have a milder flavour than the corresponding wild species. Over the years of plant cultivation, the milder varieties, that were high yielding and disease resistant, were chosen for propagation unless the plant was also used to 'spice' up other foods. Many vegetable flavours are only released from the raw vegetable when they are chopped or cooked, because the aroma compounds are tied up as glycosides (celery, lettuce) or glucosinolates (cabbage, radish), which makes them non-volatile. When the glycoside or glucosinolate linkage is broken via either enzymatic cleavage or heat, then the aroma compounds are released.

The 'green' flavour of many vegetables (peas, pepper, beans, asparagus, carrot, lettuce) comes from alkylalkoxypyrazines. Other alkylalkoxy-pyrazines are responsible for earthy aromas (potato). Phthalides give the bitter flavour to celery.

When vegetables (or fruits) are dried, many of the original flavour volatiles are removed with the water. Heat is usually used to speed the drying process (unless freeze-dried) and many of the flavour compounds change with elevated temperatures and air oxidation. New flavours can be developed from non-volatile precursors (see Maillard reaction, carotenoid degradation, lipid oxidation).

Spice Flavours

Some vegetables, such as onion and garlic, can also be considered spices. The onion is classified as *lachrymatory* as the initial flavour compound released upon enzymatic cleavage will bring tears to the eyes. Luckily it is short-lived and reacts to form other more appreciated flavour compounds. This lachrymatory compound is not formed in garlic.

Aromatic spices are the dried fruits and *aromatic* herbs are the dried leaves of plants. Volatile compounds give the characteristic aromas to the spices: eugenol (cloves), cinnamaldehyde (cinnamon) and menthol (mint). Some of these volatile substances, such as eugenol and cinnamaldehyde, also produce a slight pungent sensation via the trigeminal nerves.

The *hot* spices include chilli or red pepper, black pepper and ginger. All have aromatic characters, but the pungent sensation in the mouth is overwhelming. Garlic, nutmeg and cinnamon are also sometimes considered hot spices; however, here the trigeminal sensation occurs mainly in the nose.

In food processing, spices are often used in the form of essential oils or oleoresins. Essential oils are prepared by steam distillation of the dried ground spices and contain the volatile flavour compounds. Oleoresins are the solvent extracts of the spices and contain both the volatile essential oil as well as non-volatile resinous material and are more characteristic of the original ground spice.

Beverage Flavours

Beverage flavours can be divided into three types: unfermented, fermented and compounded. Unfermented beverages include milk and fruit and vegetable juices. Coffee might fall under this classification as it is not fermented, but because the beans are roasted to develop the flavour, it also can be considered an empyreumatic flavour.

Tea is usually classified as a fermented flavour. However, this is a misnomer. Fermentation refers to microbial growth (*e.g.* yeast), but the formation of flavour (and colour as well) during 'fermentation' in tea manufacturing is related predominantly to the oxidation of the phenolic compounds by enzymes found in the fresh tea leaves.

Alcoholic beverages use microbes to process the beverage and the chemical transformations that occur during fermentation generate flavours. However, the primary distinguishing flavours between beer and wine develop via non-fermentative processes. The bitter flavour of beer comes from hops that are transformed during the boiling of the wort before fermentation begins. Many wine flavours develop from interactions among fermentation products, flavonoid and the wooden containers during the long ageing process after fermentation has stopped.

Compounded beverage flavours can be found in the soft drinks and cordials of today that have been completely blended by flavourists. Here, the flavourist has been creative in the combination of natural and/or artificial flavours to make beverages that excite the palate.

Meat Flavours

Meats are cooked, dried, or even smoked to develop their flavours. The application of heat produces complex reactions between amino acids (often sulfur containing) and sugars (containing a carbonyl), that are given a singular name of Maillard reaction and are discussed in detail later. How long the meat is cooked, whether a dry method (broiling) or wet method (stewing) is used, and the temperatures obtained during cooking can alter the compounds formed and change the flavours dramatically.

Besides the cooking methods giving different flavour reactions, each animal contains a unique ratio of amino acids, fatty acids and sugars and thus generates its own flavours. In beef, lamb and pork, the lipids contain mostly saturated fatty acids that do not break down as quickly as do unsaturated fatty acids. However, in fish and fowl, there are many unsaturated lipids that generate flavours and small reactive molecules which interact with the amino acid/sugar reaction products to produce even more complex flavours. Also, because of these unsaturated lipids, rancid flavours develop more quickly in fish and fowl than in beef.

Fat Flavours

As suggested above, unsaturation in fats leads to oxidative cleavage and the formation of both desirable and undesirable flavours. The development of rancidity in oils is greater when oxygen and metals are present. The more refined an oil is, the quicker it develops rancidity because natural anti-oxidants, such as vitamin E, are removed during processing. When frying, the combination of heat, fat and food leads to the development of many different flavours.

Cooked Flavours

During heating in the presence of water, many flavours change (see terpenoid rearrangements) and new flavours can be developed from non-volatile precursors (see Maillard reaction; carotenoid degradation; lipid oxidation); thus these altered flavours are in a different category after they are cooked. Examples of cooked flavours can be found in soups and broths, vegetables and fruits.

Empyreumatic Flavours

Empyreal refers to being in a fire that is usually smoky and hotter than the boiling point of water. Thus, empyreumatic flavours emphasize the difference between water cooking and non-water cooking and are usually divided into three major categories: (1) smoked, (2) broiled or fried and (3) roasted, toasted or baked.

When foods such as ham are smoked, the phenolic compounds in wood transform and vaporize to infiltrate the meat and preserve it.

Broiled and fried flavours are developed at extremely high temperatures with high heat being transferred via radiation (broiling) or via conduction (through oil in frying). Examples are processed meat products and fried foods.

The roasted, toasted or baked category denotes flavours that are developed from the caramelization of sugars and deamination of amino acids. Note that in this category, the sugar caramelization products and amino acid deamination products do not always react with each other to form Maillard reaction products, as they normally do in the development of meaty flavours. Examples here are roasted coffee beans, snack foods, processed cereals and some bakery products.

Stench Flavours

Strong aromas can be produced during microbial growth, such as during the fermentation of cheeses or the spoilage of foods. Also, air oxidation can give a rancid or putrid note to foods containing unsaturated lipids, such as fish or soybeans.

CHEMICAL COMPOUNDS RESPONSIBLE FOR FLAVOUR

The many different possible flavours are due to interactions of chemical compounds with taste, trigeminal or aroma receptors. The characteristic *taste* (including trigeminal stimulations) of a food is normally related to a single class of compounds. But, an *odour* is usually elicited by a combination of volatile compounds each of which imparts its own smells. Differences in characteristics of certain aromas can be equated to the varying proportions of these volatiles. However, some substances contain trace amounts of a few volatile compounds that possess the characteristic essence of the odour. These are called *character-impact compounds*. The survey of character-impact compounds shown in Figure 1.3 is necessarily incomplete, but demonstrates the variety of chemical compound classes that elicit an aroma in the human nose.

One must also realize that the chemicals of a single compound class can elicit many diverse flavours, especially as their concentrations vary (Tables 1.2 and 1.3).

DIFFICULTIES OF FLAVOUR CHEMISTRY RESEARCH

There are many problems facing the flavourist trying to relate chemical compounds to flavours. Flavour compounds can be found in any class of chemical compounds: neutral compounds, acids, nitrogen and sulfur compounds, compounds with high volatility, compounds with low volatility, etc. These compounds are susceptible to chemical changes of various kinds. For example, aldehydes are easily oxidized to acids; amines may complex with metal ions; in the presence of acids, terpenes rearrange and isomerize; exposure to light may cause photo-oxidation or rearrangements; and polymerizations of unsaturated compounds do occur. These transformations are a real concern during the collection and concentration of foods for flavour determinations, particularly because there is such a low quantity of flavour compounds in foods. The instability of many compounds generates artefacts during the isolation of flavours, e.g. when flavours are isolated by bubbling air through oils, the triglycerides are oxidized and so new flavours are formed. Flavours also change with time and processing conditions, e.g. freshly squeezed orange juice is easily distinguished from juice reconstituted from frozen concentrate, juice that is bottled or canned and even from freshly squeezed orange juice that has been allowed to stand at room temperature open to the air for an hour or two.

It has been estimated that as few as eight molecules are required to trigger one human olfactory neuron and that as few as 40 molecules can produce an identifiable sensation.[2] Such levels are below the sensitivity limits of

FOOD	CHARACTER-IMPACT COMPOUND		COMPOUND CLASS

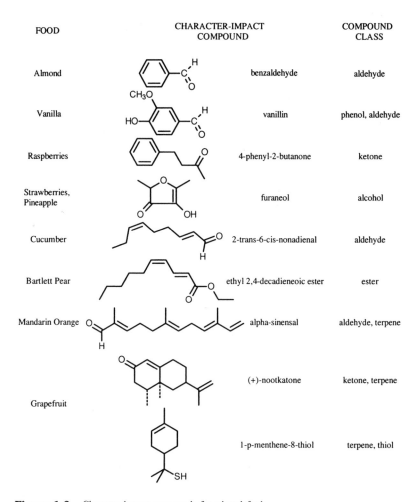

Almond		benzaldehyde	aldehyde
Vanilla		vanillin	phenol, aldehyde
Raspberries		4-phenyl-2-butanone	ketone
Strawberries, Pineapple		furaneol	alcohol
Cucumber		2-trans-6-cis-nonadienal	aldehyde
Bartlett Pear		ethyl 2,4-decadieneoic ester	ester
Mandarin Orange		alpha-sinensal	aldehyde, terpene
Grapefruit		(+)-nootkatone	ketone, terpene
		1-p-menthene-8-thiol	terpene, thiol

Figure 1.3 *Character-impact compounds for selected foods.*

present-day analytical techniques; thus, the human nose is a better detector than the best instruments of today! To give an idea of how small an amount we can smell, consider the powerful odorant, 1-*p*-menthene-8-thiol, which gives a grapefruit aroma at 10^{-4} ppb.[3] This concentration corresponds to 10^{-4} mg in one metric ton of water!

Human flavour receptors are stimulated by various compounds with different sensitivities. The flavour chemist cannot just determine the concentration of each flavour compound in foods to decide how important it is to the resulting flavour; each threshold value must also be determined. For example, the flavour thresholds of the pyrazines vary in concentration by as

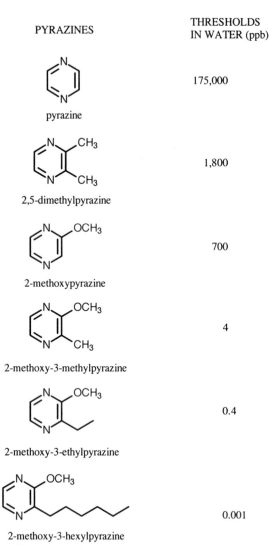

PYRAZINES	THRESHOLDS IN WATER (ppb)
pyrazine	175,000
2,5-dimethylpyrazine	1,800
2-methoxypyrazine	700
2-methoxy-3-methylpyrazine	4
2-methoxy-3-ethylpyrazine	0.4
2-methoxy-3-hexylpyrazine	0.001

Figure 1.4   *Aroma thresholds of pyrazines.*[3]

much as eight orders of magnitude (Figure 1.4).[4] The lowest level at which a compound can be detected is its detection threshold. A recognition threshold is the lowest concentration at which the flavour of the detected compound is recognized.

Since odour quality may change with concentration, a chemical such as *trans*-non-2-enal has more than one recognition threshold: just above its

Table 1.2   *Volatile compound classes and their sensory characteristics*

Compound Class	Sensory Character	Examples
Aldehydes	Fruity, green	Acetaldehyde, hexanal
	Oxidized, sweet	Decanal, vanillin
Alcohols	Bitter, medicinal	Linalool, menthol
	Piney, caramel	α-Terpineol, maltol
Esters	Fruity	Ethyl acetate, ethyl butyrate
	Citrus	Geraniol acetate
Ketones	Butter, caramel	Diacetyl, furanones
Maillard reaction products	Brown, burnt, caramel, earthy	Pyrazine(s), pyridine, furans
Phenolics	Medicinal, smokey	Phenol(s), guaiacols
Terpenoids	Citrus, piney	Limonene, pinene
	Citrus	Valencene

Table 1.3   *Non-volatile compound classes and their sensory characteristics*

Compound Class	Sensory Character	Examples
Acids		
amino	Sweet, sour, bitter	All known
organic	Sour	Citric, malic, tartaric
polyphenolic	Astringent, bitter	Chlorogenic, caffeic
Flavonoids	Astringent, bitter	Flavonols, anthocyanins
Phenolics	Medicinal, smokey	Guaicols, phenols
Sweeteners		
sugars	Sweet, body	Sucrose, glucose, fructose
high intensity	Sweet	Aspartame, acesulfame-K, saccharin, cyclamate

detection threshold of 0.1 ppb, *trans*-non-2-enal possesses a woody character. Above 8 ppb, it smells fatty, becoming unpleasant at 30 ppb, and, in an aqueous solution at 1000 ppb, it has a strong flavour of cucumber.[5] This change in flavour with increasing concentration of an individual compound can perplex the flavour chemist. Another example is 2-pentylfuran which smells beany when diluted at 1–10 ppm in oil, but when concentrated as an eluant from the gas chromatogram (GC), it elicits the distinct aroma of liquorice.[6]

Another problem that can occur is interaction when compounds are mixed together. When they interfere with flavour detection, this is called antagonism. When they enhance the ability to detect the flavour, it is called

Table 1.4   *Volatility in different media*[8]

Ketone (200 ppm)	Aqueous System*	Corn Oil System*
Acetone	10	47
Butan-2-one	14	11
Pentan-2-one	22	5.7
Hexan-2-one	29	2.7
Heptan-2-one	24	0.7

*Headspace – GC Peak Area

synergism. An example of flavour antagonism can be seen with *cis*-hex-3-enal, which has a distinct green bean aroma at 1.0 ppm in paraffin oil; when *trans,trans*-deca-2,4-dienal (12.5 ppm) is mixed into *cis*-hex-3-enal (13.2 ppm), there is almost no odour or taste.[7] Thus, the antagonist decadienal interferes with the detection of hexenal. To demonstrate flavour synergism, one can look at ketones at concentrations where each, individually, has no aroma in water (butan-2-one, 5 ppm; pentan-2-one, 5 ppm; hexan-2-one, 1 ppm; heptan-2-one, 0.5 ppm; octan-2-one, 0.2 ppm), but a solution containing all of them together at these specific concentrations has a definite aroma.[8]

Lastly, the types of compounds surrounding the flavour chemical can affect how much of that flavourant reaches the nose, tongue or walls of the mouth. For example, how well a flavourant can evaporate from the food determines its detection threshold. The more lipophilic (non-polar) a molecule, the less it will hydrogen bond to water. Therefore, it has lower solubility and higher vapour pressure in water and thus can be more easily inhaled when dispersed in water. The opposite is also true: the more lipophilic (non-polar) a molecule, the lower the vapour pressure it has in lipids and the harder it is to smell when dissolved in fats or oils. As shown in Table 1.4, different ketones placed in either water or corn oil at 200 ppm will equilibrate at different headspace concentrations above the liquid.[8] As the hydrocarbon chain is lengthened, the ketone becomes more lipophilic; less is in the air above the oil and more is in the air above the water.

OBJECTIVES OF FLAVOUR CHEMISTRY RESEARCH

Given all the difficulties expressed above, why do people accept the challenge of flavour research? Someone who is willing to untangle all the complexities of food flavours is a significant asset to the food industry. If the chemical composition of flavours is understood, then better control of these flavours is possible. If we can understand the mechanisms of their formation, then we can suppress unwanted flavours and enhance the desired ones.

During the growing and processing of foods, off-flavours occasionally develop. It is important to the quality of these foods to try to prevent or at least retard this process. Sometimes fresh flavours need to be restored to processed foods, so research is required to determine what gave the unprocessed food its original fresh flavour and what happened to this under the conditions of processing. Other times, processing creates flavours, so a thorough understanding of the reactions that produce flavour compounds can help the food processor to improve the flavour of his foods by accelerating these reactions. Occasionally, added flavour compounds not naturally found in certain foods will improve their flavours. In this way, new food items are created.

There are many different flavour-related jobs in the food industry, which include work in creation, analysis, applications and general flavour science. The creative role is very specialized. The primary focus of the creative 'flavourist' is to create flavours by combining pure chemicals or any of thousands of plant, microbial and/or animal extracts together, much like a painter mixes colours on a palette. The flavourist continuously builds on previous knowledge, as an artist, to blend flavouring materials in new ways. It is the flavour applications expert who takes that blend and applies it to a variety of different canvases. The applications expert is concerned with the flavour quality of the overall product. He or she must have a strong food science background, understanding both food chemistry and processing and their effects on foods.

The analytical flavour chemist uses instrumentation in either a quality control or a research setting. In quality control the analytical chemist may determine specifications for raw materials coming into the plant or test the flavours throughout the process to control the quality of out-going food products. In the research setting, an analytical flavour chemist might break down competitors' flavours to understand or duplicate them. He or she may even assist geneticists to breed better raw materials for food by monitoring target flavour compounds or flavour precursors.

The flavour scientist, on the other hand, broadly researches the unknown. Many flavour scientists are trying to understand how flavours interact with each other, within the food product or with the package. Outside the food industry, there are some flavour scientists who work with biologists and medical personnel to apply their methodologies to develop a more thorough understanding of the biological functions of taste and smell. For example, they might research how the variations in taste and olfactory perception influence our diet. Further, some flavour scientists are migrating from the area of flavour research in foods to semiochemicals, the chemicals used by animals and especially insects to communicate, as many of the volatile chemicals are the same.

REFERENCES

1. R.A. Flath, H. Sugisawa and R. Teranishi. In *Flavor Research: Recent Advances*, ed. R. Teranishi, R. Flath, and H. Sugisawa, Marcel Dekker, New York 1981, p. 1.
2. H. Devries and M. Stuiver. In *Sensory Communication*, ed. W.A. Rosenblith, Wiley, New York, 1961, p. 159.
3. E. Demole, P. Enggist and G. Ohloff, *Helv. Chim. Acta*, 1982, **65**, 1785.
4. R.M. Seifert, R.G. Buttery, D.G. Guadagni, D.R. Black and J.G. Harris, *J. Agric. Food Chem.*, 1970, **18**, 246.
5. D.A. Forss. In *Flavor Research: Recent Advances* eds., R. Teranishi, R. Flath and H. Sugisawa, Marcel Dekker, New York, 1981, p. 125.
6. S.S. Chang, T.H. Smouse, R.G. Krishnamurthy, B.D. Mookherjee and R.B. Reddy, *Chem. & Ind.*, 1966, 1926.
7. P.W. Meijboom, *J. Am. Oil Chem. Soc.*, 1964, **41**, 326.
8. W.W Nawar and I.S. Fagerson, *Food Technol.*, 1962, **16**(11), 107.

Chapter 2

Flavour Compounds

CHEMICAL COMPOUND CLASSES AND THEIR FLAVOUR RESPONSES

A single class of chemical compounds can elicit many different flavours, but normally evoke one type of response from one type of receptor. That is to say, a compound class, such as terpenes, is usually volatile and elicits an aroma. However, the functional groups that are on the terpene as well as the overall size and shape of the molecule determine its unique flavour. Most sugars are sweet (or bitter), and most esters are fruity. These generalities are shown in Tables 1.2 and 1.3.

The first major classification of flavour chemicals is by volatility or how easily the compound evaporates into the air. This is a good working classification because the volatile compounds travel through air into the nose or the oesophageal passageway to arrive at the nasal receptors and elicit a response. The non-volatile compounds must be carried to the taste buds of the tongue or the inner lining of the mouth via food or saliva to elicit a response.

Volatile Flavour Compounds

It has been estimated that humans are responsive to 5000 to 10 000 aromatic compounds.[1,2] So far, over 2600 chemicals have been recognized as volatile components of aromas.[1,3] They are partially lipid-soluble organic chemicals of low molecular weight (below 300 m.u.). Many of these have been identified since the invention of gas chromatography (GC). However, an analytical flavour chemist cannot simply identify and quantify the volatile compounds in foods, as the GC analysis does not establish the contribution of individual volatile compounds to flavour. It is now well established that many of the larger peaks on gas chromatograms of foods do not correlate to flavour. For example, limonene is the major component by weight of citrus oils, but it has a weak aroma. It is the oxygenated terpenes present in small amounts in these oils that have the major impact on the flavour. Although hydrocarbons like limonene may not have much aroma, they do act as a solvent for the powerful odorants. A volatile solvent might allow odorants to evaporate more efficiently (perhaps as an azeotrope) and help to carry them through the air to the nose, such as steam does when that tempting soup is simmering.

What follows are examples of the range of aroma characteristics provided by individual chemical groups. It must be remembered that a single fruit, vegetable or food contains a variety of aroma chemicals and that some chemicals can contain more than one of these functional groups. Thus, the flavour characteristics of each functional group are, by necessity, a generalization and may not apply to every member of that group.

Aldehydes. Aldehydes play an important role in providing flavour characteristics of a wide range of foods. The unsaturated aliphatic aldehydes tend to produce stronger aromas than their saturated analogues.

In vegetables, for example, the 'green' flavour note is produced by *trans*-hex-3-enal and in leaves it is mainly *trans*-hex-2-enal. The overall flavour of dehydrated potato products is believed to be determined by a number of alk-2-enals found together with their corresponding saturated analogues. Short-chain saturated aliphatic aldehydes constitute an essential part of the natural tomato flavour. Cucumber flavour comes from (E,Z)-nona-2,6-dienal and (E)-non-2-enal. Present at only 0.9 ppb (45 times its threshold), (Z)-non-6-enal gives musk melon its typical aroma. In beer, 0.5 ppb of (E)-non-2-enal is perceived as a strong papery-stale aroma.

Aldehydes are also responsible for many fruit aromas. Some flavour differences in varieties of apples were found to be a consequence of quantitative variations in the formation of *trans*-hex-2-enal and *trans*-hex-3-enal as well as of the corresponding alcohols. Benzaldehyde is strongly reminiscent of almonds and is associated with cherry flavour. The odour of citrus comes from aliphatic aldehydes, like decanal, having 8–10 carbon atoms and oxygenated terpenes, such as terpineol and citral.

The off-flavour component of partially hydrogenated vegetable oils and dried milk comes from (E)-non-6-enal. (E,Z,Z)-Deca-2,4,7-trienal and (E,E,Z)-deca-2,4,7-trienal are formed by oxidation of linolenic acid and are responsible for the formation of the 'fishy' off-flavour in rancid mackerel oil. The rancid off-flavour in frozen stored cod is attributed to *trans*-hept-4-enal. At concentrations above 2 ppm, *trans*-hept-4-enal produces a rancid note in butter and soybean oil. But at a level of 1 ppb, *trans*-hept-4-enal confers a creamy flavour to deodorized butter.

The *cis*- and *trans*-deca-2,4-dienals have been found in a wide variety of foods, including cottonseed oil, soybean oil, milk fat, beef fat, cooked chicken, tomatoes and rye crisp bread, and are believed to impart a desirable deep-fat flavour.

Alcohols. The odour threshold of alcohols is considerably higher than that of the corresponding aldehydes, so they are normally less important to flavour profiles. For example, as the storage time of fruit preserves or jellies is prolonged, the carbonyl content (aldehydes and ketones) decreases due to the formation of corresponding alcohols; this conversion is equivalent to a decrease of the flavour value.

trans-Hex-3-enol has a green note and occurs in a wide range of higher plants in conjunction with *trans*-hex-3-enal. *trans*-Hex-3-enol was first discovered in fermented tea leaves in 1895.

The main component of the volatile constituents of yellow passion fruit is hexan-1-ol. However, the flavour character is not determined by this alcohol, but by trace amounts of sulfur-containing compounds.

Ketones. Although most ketones have relatively high flavour thresholds, some, like the alkan-2-ones of different chain length in dairy foods, are still important to flavour. A compound responsible for a 'metallic' flavour in oxidized butter was shown to be oct-1-en-3-one. A strong blue cheese flavour is imparted to dairy products by the addition of non-8-en-2-one.

Acids. Besides producing a sour taste in the mouth, the volatile acids also can impart an aroma. Butyric and caproic acids interact at individual sub-threshold concentrations to produce a desirable butter flavour. An essential flavour component of Swiss cheese is propionic acid. Both hex-2-enoic and hex-3-enoic acids are important in raspberry flavour. When animal and vegetable oils are heated in the presence of air, alk-2-enoic and alk-3-enoic acids are formed and become an important component of fried food flavour.

Several alkyl-branched fatty acids are powerful components of food flavours. An important constituent of the aroma of cranberry is 2-methylbutyric acid. Contributing to the distinct aroma of Turkish tobacco smoke are 2-methylvaleric acid and isovaleric acid. Isovaleric acid has the lowest flavour threshold (0.7 ppm) of all saturated fatty acids and is reminiscent of Limburger cheese. The undesirable odour of mutton is attributed to branched chain and unsaturated fatty acids having 8–10 carbon atoms. (Z)-2-Methyl-pent-3-enoic acid is used in aroma compositions to impart a sweet, green sharp strawberry character.

Esters. Aliphatic esters in various combinations play a major role in many fruit flavours. Hexyl acetate gives Cox's Orange Pippin apples their strong characteristic aroma. Much of the flavour of Golden Delicious apples comes from hexyl 2-methylbutyrate. Pineapple flavours contain the methyl and ethyl esters of (E)-hex-3-enoic, (Z)-dec-4-enoic and (E)- and (Z)-oct-4-enoic acids. Ethyl (E,Z)-deca-2,4-dienoate is the character-impact compound of

Bartlett pears. Methyl- and ethyl-cinnamates provide sweet, honey notes and are associated with strawberry flavour.

Lactones. Due to their mostly low odour threshold values, averaging around 0.1 ppm, lactones have a high flavour value. Lactones are internally formed esters and in chemical equilibrium with their corresponding acids; 4-hydroxy acids transform into γ-lactones and 5-hydroxy acids transform into δ-lactones. Thus, during storage, the sweet, pineapple aroma of ethyl 5-hydroxyoctanoate and ethyl 5-hydroxydecanoate, changes into an intense coconut-like aroma as these hydroxy esters cleave to expose the carboxylic acid group and cyclize into their corresponding δ-lactones.

In many dairy products, δ-lactones impart the distinct impression of sweet cream and milk. A butter-like note is imparted by δ-decalactone. Most δ-lactones are found in animal products, whereas γ-lactones preferentially occur in plants.

Various γ-butyrolactones have been identified in the volatiles of many foodstuffs (chicken, beef, apricot, passion fruit, raspberry, strawberry, raisins, plums, tomatoes, pineapple, citrus, coffee, tea, roasted onions, popcorn, roasted peanuts, cocoa, bread, beer, wine, vinegar and mushrooms), giving many individual flavours but especially a sweet roasted note. Peach flavour contains many lactones: δ-decalactone, γ-hexalactone, γ-heptalactone, γ-octalactone, γ-decalactone, and γ-dodecalactone.

Furans. Occurring in the volatiles of all heated foods, oxygenated furans are usually formed from carbohydrates in the Maillard reaction. Furfurals and furanones generally impart caramel-like, sweet, fruity characteristics to foods. Described as burnt and caramel-like, 5-methylfurfural also produces a slight meaty flavour. A rum-like aroma comes from 1-(2-furyl)propan-2-one. Found in beef broth, pineapples and strawberries, 2,5-dimethyl-4-hydroxyfuran-3(2H)-one has a caramel-like, burnt pineapple-like aroma, which at low concentrations changes to a hint of strawberry.

Phenols. Ethyl-, vinyl- and methoxy-phenols are considered to be important contributors to off-odours in wine. Since volatile phenols are generally not present in grape juices, they must arise from the metabolism of some precursors. Two different pathways are involved: 1) the biochemical degradation of phenolic acids during yeast fermentation and 2) the chemical degradation of lignin from the storage barrels.

Lignin also provides phenols in smoked products via the slow burning of wood and makes a major contribution to the aroma of cured meats. Thus, smoked bacon and ham contain phenols and guaiacols (methoxyphenols) which are responsible for their characteristic smoke flavour.[4]

Many character-impact compounds of spices are phenols; for example, eugenol from cloves (both an alcohol and a phenol) and vanillin from vanilla (both an aldehyde and a phenol).

Terpenoids. Terpenoids are substances derived in nature from the metabolic intermediate, mevalonic acid, which provides the basic structural unit, the isoprene unit. Hemi-, mono-, sesqui-, and di-terpenoids, having one, two, three or four isoprene units, respectively, are well known, but it is the monoterpenoids that provide the character-impact flavours of many herbs, spices and citrus fruits. The terpenes (hydrocarbons) are found in the essential oils of most plants, but have little flavour of their own. Usually the oxygenated terpenes have flavour threshold values much lower than those of the hydrocarbons. The hydrocarbon terpenes may simply interact with the oxygenated terpenes as a solvent to enhance the ability of the flavour compounds to reach the organoleptic receptors.

Linalool, geraniol, nerol and citronellol are the monoterpene oxygen-containing derivatives that occur most abundantly in nature. Their esters and aldehydes are also common. All the natural compounds derived from linalool and citronellol exist in optically active forms of both enantiomeric series. The optically active forms differ in their odour qualities. (−)-Linalool is a component of basil aroma, whereas its enantiomer, (+)-linalool, is a component of volatiles from coriander. A more oxygenated derivative of linalool, dimethylocta-1,5,7-trien-3-ol, having an aroma similar to that of lime-tree blossoms, has been found in various grapes, wines, and teas.

Limonene is the most typical representative of the monocyclic monoterpenes and can take two optically active forms, with (+)-limonene found in citrus species and (−)-limonene predominating in *Mentha* and *Pinus* species. Most enantiomers differ in their odour qualities. For example, the (+)-enantiomer of carvone imparts its odour to caraway, whereas the impression of spearmint is produced by (−)-carvone. However, not all people can perceive this difference.

The occurrence of bicyclic monoterpenes, such as α-pinene, β-pinene, camphene and 3-carene, in plant volatiles is both diverse and abundant. Oxygenated bicyclic monoterpenes, such as camphor, impart similar aromas (usually designated as camphoraceous) because of their spherical structure.

Compared to the monoterpene derivatives, the quantity of the larger sesquiterpene compounds (made up of three isoprenoid units) in essential oils is small, but very important to flavour; some of their representatives have been found to be character-impact compounds. Caryophyllene oxide, a marijuana component, is responsible for the recognition of *Cannabis sativa* by dogs trained to find the drug. These dogs can detect as little as 1 μg of the epoxide which puts their detection threshold at about 10^6 times lower than in man. The humulene epoxides contribute to the hop character in beers. (+)-β-Bisabolol and bisabolene oxide found in buds of the cotton plant have an apple blossom-like odour and are quite attractive to the boll weevil.

Compounds having the bisabolane skeleton, (−)-zingiberene and (−)-ses-quiphellandrene, are present in ginger oil and provide the warm, woody, spicy, sweet, tenacious aroma representative of ginger ale.

The grapefruit-like compound, nootkatone, illustrates again how enan-tiomeric forms differ in both odour profiles and odour strength. The threshold concentration of (+)-nootkatone varies from that of (−)-nootkatone by a factor of 1000 in water and by as much as 2000 in the vapour phase.

Sulfur-containing compounds. The sulfur group of any compound has a high flavour impact because it binds strongly to the olfactory receptors. For example, thiols have dramatically lower odour thresholds than their analogous alcohols. The potent character-donating constituent of grapefruit, 1-*p*-menthene-8-thiol, a sulfur-containing terpene, has the low threshold of 10^{-7} ppm.

Furans with thiol, methylthio and disulfide substituents have been shown to have meaty characteristics. 2-Methylfuran-3-thiol, its methylated deriva-tive, 2-methyl-3-(methylthio)furan and its disulfide, bis-(2-methyl-3-furyl)disulfide are generated during the cooking of beef. The threshold value of the disulfide has been reported as 2×10^{-8} ppm, one of the lowest known threshold values. The thiol also has a very low threshold value, estimated to be 2500 times lower than that of its methylated derivative.

At high concentrations, many sulfur compounds generate a stench aroma. These have been used to detect gas leaks; odourless natural gas is required by law to contain an odorant, such as tributylmercaptan or ethylmercaptan, so that people will recognize when the concentration of natural gas has reached dangerous levels.

The group of isothiocyanates that give the pungent aroma in cruciferous plants, such as horse-radish, mustard, kohl-rabi and cabbage, appear during mechanical cell injury from non-volatile precursors. The generation of aromas following cell damage also occurs in the allium plants, such as onion, garlic and chives. For example, the thiosulfinates, thiosulfonates, disulfides and trisulfides of allium species all arise from sulfenic acids generated enzymatically from cysteine sulfoxides during chopping, blending or other-wise mechanically injuring the cell walls.

Dimethyl sulfide is the principal volatile flavour component of processed tomatoes (but is absent in fresh tomato) and is formed during the heating process from an *S*-methylmethionine sulfonium salt. Other sulfur com-pounds which have been characterized in heat-processed tomato are methional, 2-methylthioethanol, 3-methylthiopropanol and 2-methyl-thioacetaldehyde. Strongly associated with fresh tomato flavour, 2-isobutyl-thiazole is a good example of the effect of media on flavour perception. In water, it has a spoiled vine-like, slightly horse-radish-type flavour; when added to canned tomato juice, it produces a more intense, fresh tomato-like flavour and improves the mouthfeel of the juice.

Trace amounts of hydrogen sulfide (H_2S) are necessary for mint extracts to smell right. However, it is missed in many GC analyses because it is so volatile that it comes off with the solvent peak.

Non-Volatile Flavour Compounds

Only a few volatile compounds are detected by the taste buds of the tongue or the inner lining of the mouth. Most receptors in the mouth elicit a response from non-volatile compounds, which must be carried physically via food or saliva to these receptors. Non-volatile compounds, by definition, cannot have an aroma, but they can be broken down to release volatile compounds, such as by the hydrolysis of terpene glycosides or the enzymatic cleavage of glucosinolates.

Amino acids and peptides. The dipeptide, aspartame (*N*-L-aspartyl-L-phenylalanine methyl ester), is a commercial low calorie sweetener that is 180 times sweeter than sucrose by weight. It has a clean, sweet taste without a bitter or metallic after-taste.

The amino acids alanine and proline are sweet but with a detectable bitter note. The amino acids phenylalanine, isoleucine and tryptophan are bitter. Amino acids and/or peptides formed from the degradation of milk proteins are primarily responsible for the bitterness found in dairy products.

Organic acids. Citric acid gives the tartness to citrus fruits such as oranges and lemons. Citric acid and malic acids give the fresh tartness to apples and tomatoes. In grapes, tartaric and malic acids are the major ones. Dairy products such as yogurt and buttermilk contain lactic acid.

Sugars. In green plants, the main carbohydrate which is translocated from one part of a plant to another is the disaccharide, sucrose. In ripening fruits, sucrose is rapidly converted by an invertase enzyme into glucose and fructose as soon as it enters the fruit. In fruits such as tomatoes, only 5–8% of the fresh weight is solids, and half of this is sugars. In storage roots such as potatoes, the sucrose has been converted into starch.

In citrus juices, sugars found in significant quantities are sucrose, glucose and fructose; in orange juice, they occur in the approximate ratio of 2:1:1. The total solids content (sugars represent 63–80%) and the ratio of total sugars to total acids present in citrus juice are so important to the desirable flavour that these values are the primary criteria for determining legal maturity for oranges, grapefruit and tangerines in Florida.

Salts. Salts are ionic compounds that dissociate in water into anionic (−) and cationic (+) species. Cations are primarily responsible for salty taste, with sodium having the lowest threshold. However, table salt (sodium chloride) tastes saltier than other sodium salts, so the anion also exerts an effect on the taste buds (see Chapter 3). Potassium chloride is sometimes used as a salt replacer in low sodium foods, but it imparts a bitter flavour as well as being salty.

Monosodium glutamate (MSG) and the 5'-nucleotides supply the characteristic savoury quality of the 'umami' taste. However, their more important sensory contribution is their flavour-enhancing property which only occurs within a limited pH range (5.5–8.0) in most food systems. It has been suggested that our sensory receptors for flavour enhancement are proteinaceous in nature and sterically different from the basic taste receptors. That is, L-glutamate binds preferentially to specific taste receptors and that the presence of certain 5'-nucleotides significantly increases the level of L-glutamate interactions at the receptor surface, probably due to conformational changes in the protein.[5]

The most important meat flavour enhancers are MSG and inosine 5'-monophosphate (IMP). IMP is formed during ageing (beef produced in the US is normally aged for at least seven days) from adenosine-5'-triphosphate (ATP). It is the major nucleotide in dead muscle and is more important as an enhancer in natural meat than the closely related guanosine-5'-monophosphate (GMP), despite the fact that GMP is the more efficient potentiator. These nucleotides exhibit synergism with each other and certain amino acids (glycine) or dipeptides (asparginine-L-aspartate). The 5'-nucleotides enhance mouth-filling, meaty, brothy, dry and astringent qualities; they suppress sulfurous and hydrolysed vegetable protein notes while not affecting sour, sweet, oily, fatty, starchy and burnt qualities.

Salts of heavy metals (*e.g.* lead acetate) are sweet, hence the attraction of lead-based paints to children.

Flavonoids and Alkaloids. Most flavonoids and alkaloids are bitter and/or astringent. Both flavonoids and alkaloids contain heterocyclic ring systems, with oxygen being the heteroatom in flavonoids and a basic nitrogen atom being the heteroatom in alkaloids.

Alkaloids such as nicotine in tobacco and caffeine in coffee and tea are bitter. Naringin, along with limonin, gives grapefruit its refreshing bitterness. It is the flavonoids or, more specifically, the catechins that are the predominant flavour components in tea and provide both astringency and bitterness. One of the main differences between green and black tea is the size of the catechins. Both types of product can come from the same tea plant; the monomeric catechins from the growing plant are present in green tea, whereas the dimers, trimers and larger polyphenols of catechins arise from enzymatic polymerization during the 'fermentation' of leaves to produce black tea.

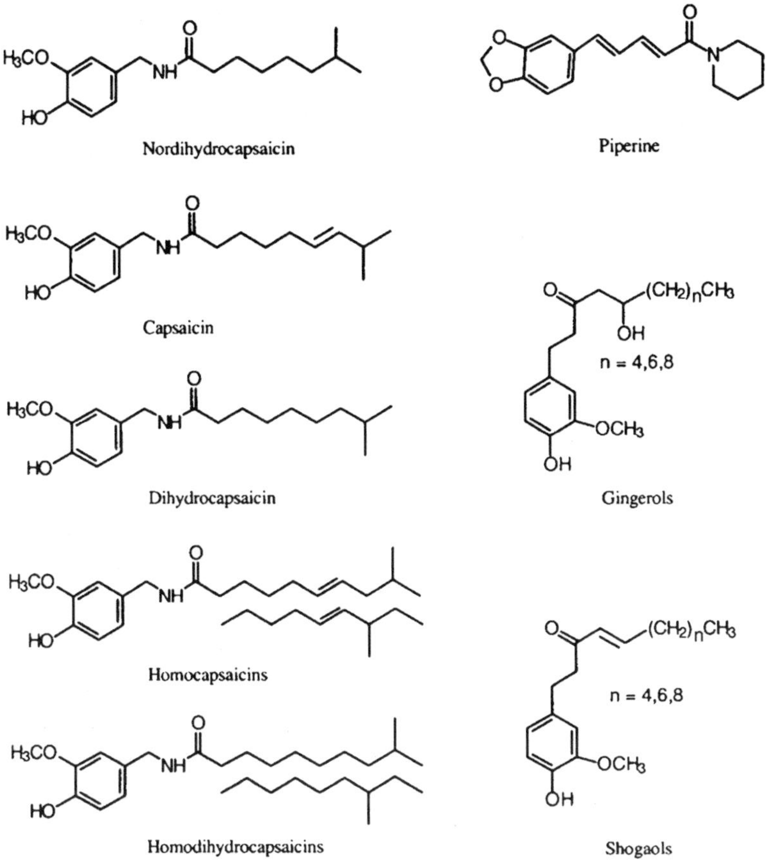

Figure 2.1 *Pungent compounds found in capsicums, black pepper, and ginger.*

Bitterness and astringency in wines are generally attributed to polymeric flavonoid phenols. However, the monomeric flavonoids are more bitter than astringent. As the size of the flavonoid polymer increases during ageing of wine, astringency increases relative to bitterness.

Apples high in tannins (total polyphenolic content) usually produce cider that has a problem with bitterness and astringency. While there are many different types of polyphenols in apples, only procyanidins (polymeric flavan-3-ols) have any significant bitterness and astringency at the levels in which they are found in cider.

Phenols. The compounds responsible for the well-known hot sensation of chilli peppers belong to a family of vanillylamides known as the capsaicinoids (Figure 2.1). These consist of a phenol, an amide linkage and a long-chain fatty acid of variable lengths. All three of these segments are necessary for the trigeminal nerve endings to be stimulated; that is, a compound must have all three chemical moieties to be perceived as hot.

Large or frequent doses of capsaicinoids reduce the sensitivity of an individual to subsequent usage (adaptation). This has caused problems with reproducibility in sensory evaluations.[6] Capsaicinoids also affect thermoregulation – body temperature actually drops with small doses. This hypothermia is associated with increased perspiration,[7] which is why hot foods are usually eaten in countries where the average temperatures are too high for comfort. Large doses of capsaicinoids, on the other hand, can cause desensitization and thus allow overheating.[7].

When Columbus arrived in America looking for the spices of the Orient, he named the pungent fruit of a small annual shrub (*Capsicum*) after the pungent berries of a vine (*Piper niugrum* L.). The pungency in black pepper is due to piperine, which has a somewhat different structure from the capsaicinoids. It is not strictly a phenol, but instead contains a methyl-enedioxy group attached to the aromatic ring. Because of this and the movement of the hydrogen-bonding group away from the aromatic ring, piperine is much less pungent than the capsaicinoids, but gives a similar type of warming sensation.

The two families of phenols, gingerols and shogaols (Figure 2.1), are responsible for the pungency of ginger and have similar structures. In fresh ginger, gingerols are the prominent species, while in commercial oleoresins, the shogaols are the major pungent phenols. The taste threshold pungencies of [6]-gingerol (17 ppm) and [6]-shogaol (7 ppm) are comparable to that of piperine (10 ppm), which is about 100 times less pungent than the capsaicinoids (0.06 ppm). The higher homologues, [8]- and [10]-gingerols and shogaols, have a lower pungency which follows the same pattern as found among the capsaicinoids.

Shogaol and zingerone are formed from gingerol during processing and storage, or by distillation, even under reduced pressures (Figure 2.2). With formation of shogaol by the dehydration of gingerol, the pungency doubles – which is why aged ginger usually has a stronger bite. However, when gingerol transforms via the retro-aldol reaction into zingerone (sweet–spicy, heavy floral, vanilla-like) and hexanal (off-flavour) the pungency is completely lost because all three chemical moieties, phenol, hydrogen bonding group in the middle and hydrophobic group at the opposite end, are necessary for the trigeminal nerve endings to be stimulated.

Isothiocyanates. Giving the pungent flavour to cruciferous plants, isothiocyanates are quite evident when one bites into a radish (4-methylthiobut-3-enyl isothiocyanate, $CH_3S–CH{=}CH–CH_2CH_2NCS$) or horse-radish (allyl isothiocyanate, $CH_2{=}CHCH_2NCS$). The compounds responsible for the bite in prepared mustards are allyl isothiocyanate (black mustard) and *p*-

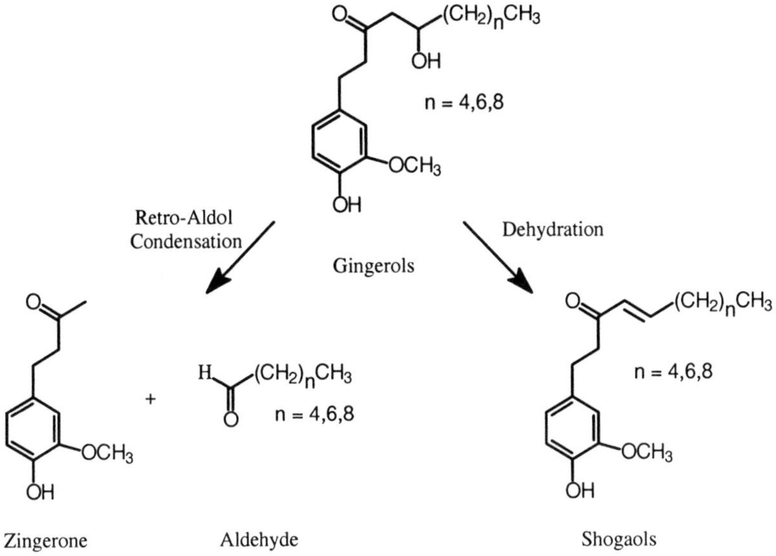

Figure 2.2 *Degradation of gingerols.*

hydroxybenzyl isothiocyanate (white mustard). The smaller molecules are volatile and generate the trigeminal response in the nose, whereas the larger non-volatile isothiocyanates produce this sensation in the mouth.

FLAVOUR DEVELOPMENT DURING BIOGENESIS

Many flavours, especially those in fruits and vegetables, are the secondary products of various metabolic pathways. Secondary metabolites can be distinguished functionally from primary metabolites, such as proteins, carbohydrates and lipids, in that they do not seem to have any direct physiological function. Secondary metabolites act as an interface between the producing organism and its environment; they may be produced to combat infectious diseases, to attract pollinators and to discourage or encourage herbivores. Secondary metabolites are often synthesized from primary metabolites. For example, alkaloids, flavonoids, phenols and other aromatic compounds can be synthesized from the amino acid primary metabolites phenylalanine, tyrosine and tryptophan.

Fruit Flavours

All fruits share a very high proportion of the same volatile compounds. For example, of the 17 esters identified in banana volatiles, only five are not found in apples. Most volatile constituents in fruits contain aliphatic hydrocarbon chains, or their derivatives (esters, alcohols, acids, aldehydes, ketones, lactones), with saturated ones predominating in apples, unsaturated ones predominating in pears and branched chains predominating in bananas. These constituents can be viewed as ripening products which develop from two different sources: fatty acids (described under lipid metabolism – β-oxidation) and amino acids (described under amino acid metabolism).

Esters are by far the largest chemical category of volatiles from fruits. The apple, which has a prominent odour, largely produces and emits esters of relatively low molecular weight. Pear odours are more subtle and contain esters of higher molecular weight.

Fruits are *climacteric* if their ripening stage is triggered by increased levels of ethylene. In climacteric fruits, ripening has also been shown to be triggered by vapours of acetic, propanoic and butanoic acids, probably by enzymatic transformation of these acids into ethylene. These fruits have living respiring systems, even after they have been harvested. If the air is depleted of much of its oxygen and enriched in carbon dioxide, respiration, and thus ripening, is slowed. These facts have been utilized in the controlled-atmosphere storage of apples and pears; a low oxygen atmosphere surrounds the fruits in cold storage and is continuously filtered through charcoal to absorb the volatiles given off by the fruits. This increases their storage time so that we can enjoy fresh fruits the year around. Bananas are also picked green and their ripening hastened with ethylene during transport.

The acidity in fruits is produced by the accumulation of the two citric acid cycle acids, citric and malic. Ripening is associated with a rapid decrease in the enzymes of the citric acid cycle; although more than half of the enzyme activity is lost, citrate actually increases at this time and malic decreases. This has been linked to the production of an nicotinamide-adenine dinucleotide phosphate (NADP)-linked enzyme which decarboxylates malic acid into pyruvate. The increased concentration of pyruvate induces citrate synthase to produce more citrate. This contributes to the increased respiration of carbon dioxide by fruits at the climacteric stage.

Vegetable Aromas

Most vegetable aromas develop during cellular disruption, which releases enzymes that act upon non-volatile precursors. Before the cell walls of the vegetable are broken, there is no aroma; the aroma compounds are chemically bound to precursor handles such as sugars (*e.g.* terpene glycosides) and amino acids (*e.g.* cysteine sulfoxides).

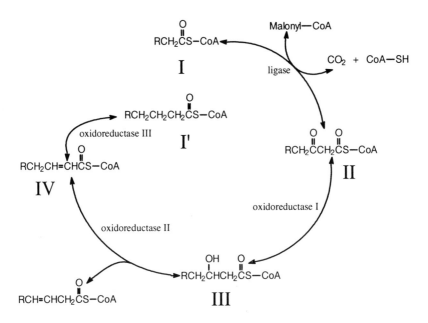

Figure 2.3 *Fatty acid spiral in plants.*

Lipid Metabolism

β-*Oxidation pathway.* Natural plant volatiles, such as aliphatic esters, alcohols, acids and carbonyls, can be derived from fatty acid metabolism. Most unripe fruits, *e.g.* apples, bananas and strawberries, produce a variety of fatty acids which, during ripening, are converted into esters, ketones and alcohols. Accordingly, via the reversible fatty acid spiral (Figure 2.3), compound I affords saturated fatty acids, II and III yield methyl ketones and secondary alcohols and IV can furnish unsaturated fatty acids, unsaturated aldehydes, unsaturated alcohols, and eventually γ-lactones.

For example, aroma differences in apple varieties are related to the proportion of alcohols and esters which depends upon competing reaction rates in the β-oxidation pathway of the various fatty acids. The enzyme complex catalysing β-oxidation is located in the mitochondria and is inactivated during disruption of plant cells.

Oxidation via lipoxygenase. During the cell break-down that occurs naturally in fruit ripening and especially when vegetative cells are mechanically broken, the oxidation of unsaturated fatty acids is catalysed by lipoxygenase (Figure 2.4). This enzyme exists in multiple forms (four in soybeans, two in corn) and converts polyunsaturated fatty acids, mostly linoleic and linolenic acids, into

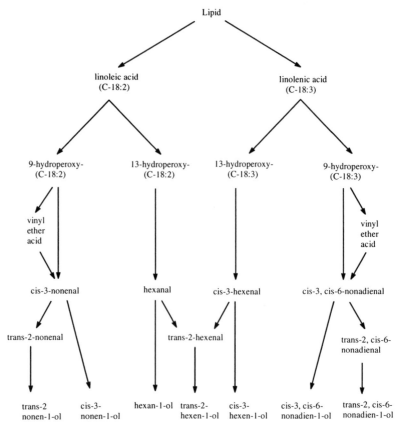

Figure 2.4   *Oxidation via lipoxygenase.*

hydroperoxides. These break down to give aldehydes, alcohols and esters (usually with the help of other enzymes – hydroperoxide lyases, alcohol dehydrogenases, isomerases and esterases). Each type of plant has its own set of lipoxygenases, pH and medium conditions. These factors influence the path of the break-down, resulting in different volatile products from the same polyunsaturated fatty acid starting material.

Lactone formation. When the lipid oxidation described above forms 4- or 5-hydroxy acids, lactones (internal esters) are usually formed which stabilize the hydroxy fatty acid so further oxidation does not occur.

Amino Acid Metabolism

Banana and/or apple aromas. The biosynthesis of branched chain alcohols and esters in fruits and vegetables is shown in Figure 2.5. Enzymes remove both

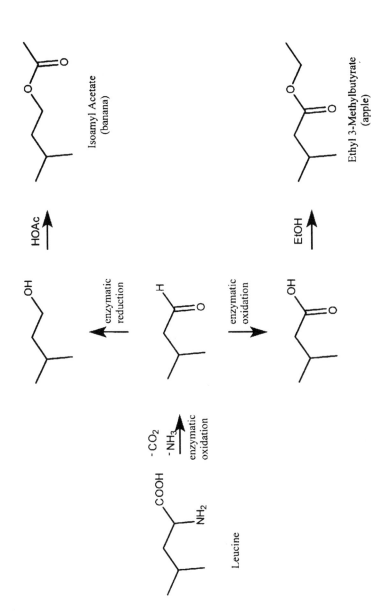

Figure 2.5 *Enzymatic conversion of leucine into flavour volatiles.*

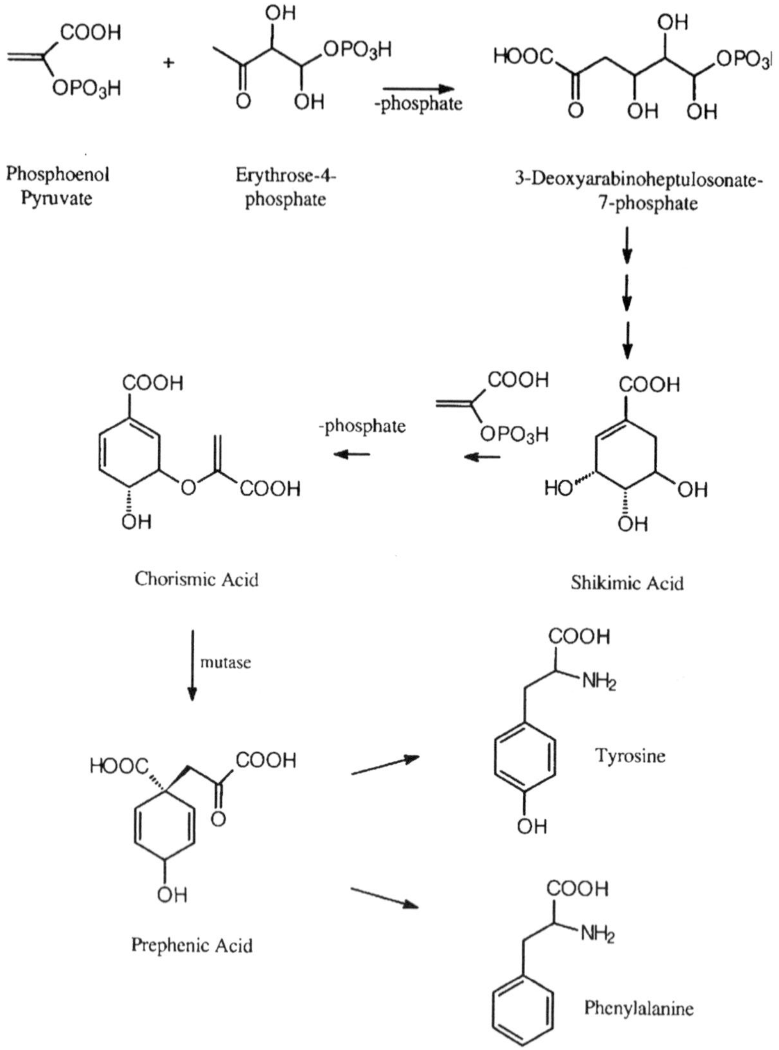

Figure 2.6 *Shikimic acid pathway to the production of aromatic amino acids.*

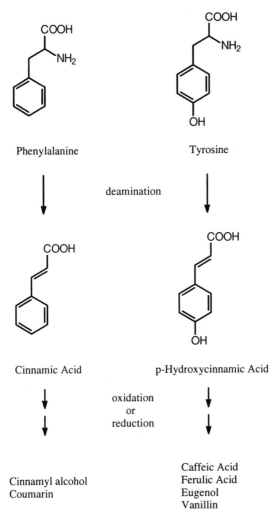

Figure 2.7 *Flavour compounds from aromatic amino acids.*

amine and carboxyl groups from the amino acid to produce aldehydes. The aldehydes are either enzymatically oxidized or reduced and then esterified. *Shikimic acid pathway.* The aromatic amino acids, phenylalanine and tyrosine, are formed by the shikimic acid pathway (Figure 2.6). Many aromatic flavour compounds from spices, such as eugenol (cloves), cinnamaldehyde and coumarin (cinnamon), are generated from these amino acids via deamination and oxidation or reduction (Figure 2.7).

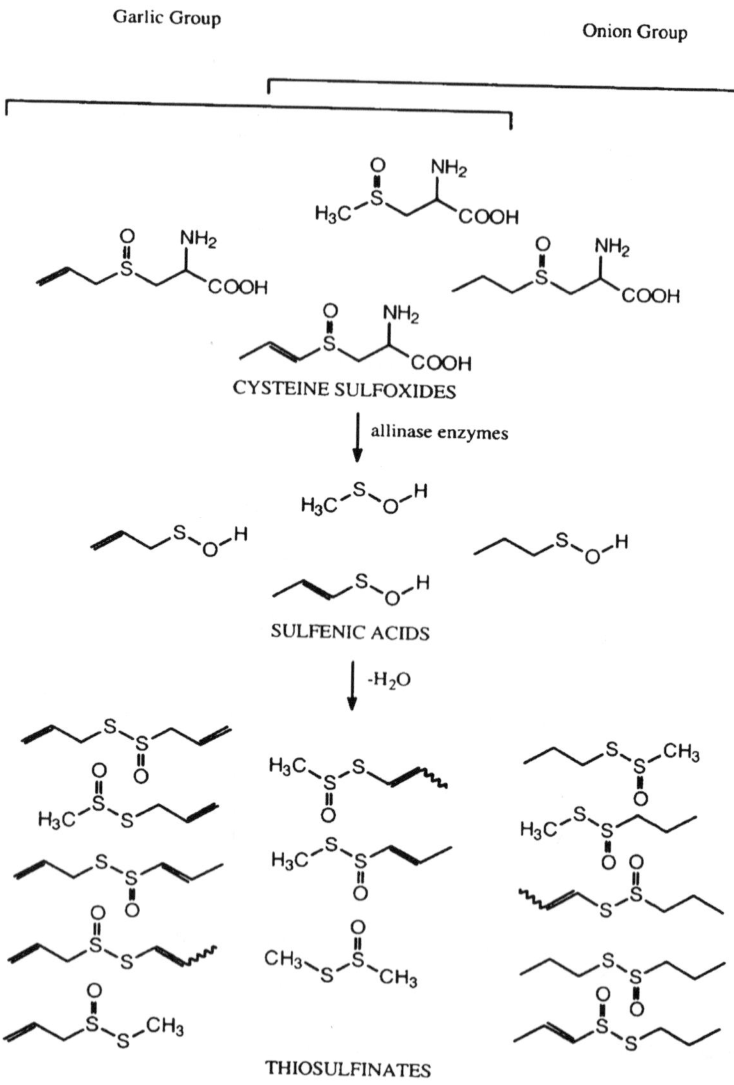

Figure 2.8 *Enzymatic flavour development in onion and garlic.*

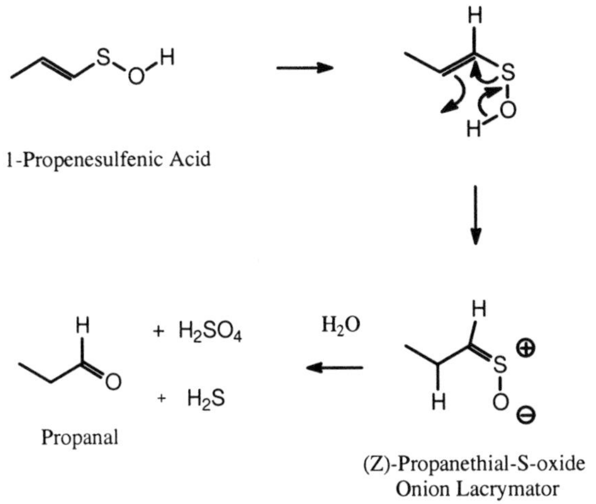

1-Propenesulfenic Acid

+ H₂SO₄

+ H₂S

Propanal

H₂O

(Z)-Propanethial-S-oxide
Onion Lacrymator

Figure 2.9 *Rearrangement of sulfenic acid, formed enzymatically into the onion lachrymator.*

The initial step to making shikimic acid is the condensation of phosphoenol pyruvate, from glycolysis, and erythrose 4-phosphate from the pentose phosphate pathway. After cyclization, dehydration, and reduction, shikimic acid is formed. Condensation with a second molecule of phosphoenol pyruvate gives chorismic acid, which is the precursor of the aromatic amino acids phenylalanine and tyrosine.

Fresh onion and / or garlic aromas. Enzymatic flavour propagation, during which the initial reaction products are unstable and undergo further reaction, is utilized by the allium family of vegetables which include onion (*Allium cepa*), garlic (*Allium sativum*), chive (*Allium schoenoprasum* L.), shallot (*Allium ascalonicum* auct.) and leek (*Allium porrum* L.). The important flavour precursors (or enzyme substrates) are amino acid sulfoxides. Cutting or crushing allium plants releases allinase enzymes that convert the sulfoxide precursors into intermediate sulfenic acids, which then either rearrange to form the lachrymator (*Z*)-propanethial-*S*-oxide, or condense to form odorous thiosulfinates (Figure 2.8). The lachrymator and mixed thiosulfinates are the compounds primarily responsible for the characteristic flavour of *freshly cut* members of the genus *Allium*.

Onion and leek contain mainly 1-propenyl-cysteine sulfoxide and 1-propenyl-γ-glutamyl sulfoxide, whereas the major flavour precursor in chives is methylcysteine sulfoxide and methyl γ-glutamyl sulfoxide. When 1-propenylsulfenic acid is formed, it rearranges spontaneously to become the lachrymator (*Z*)-propanethial-*S*-oxide (Figure 2.9).

The allyl group is absent in onion, scallion, shallot, leek and chive, whereas it is the major sulfur-containing group in garlic. Because allylsulfenic acid

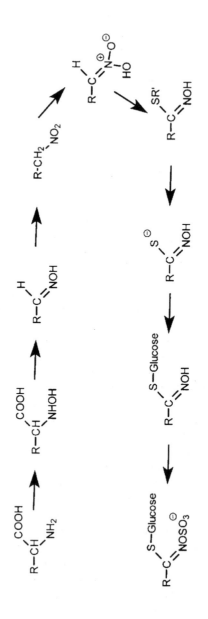

Figure 2.10 *Biosynthetic pathway to glucosinolates.*

does not rearrange, garlic does not cause tears and allicin (diallylthiosulfinate) is produced when it is cut. The maximum amount of diallylthiosulfinate is formed 30 seconds after the allinases are released and decreases in concentration until five minutes later, when the mixed allyl methyl thiosulfinates are at their maximum.[8] The thiosulfinates containing propenyl groups are formed ten times more rapidly than thiosulfinates containing methyl groups. Because the asymmetric thiosulfinates in garlic, $MeS(O)SCH_2CH{=}CH_2$ and $MeSS(O)CH_2CH{=}CH_2$ are typically found in a 2:1 ratio, a mechanism has been postulated by which the more slowly formed methanesulfenic acid reacts with diallylthiosulfinate rather than allylsulfenic acid.[7]

S-Methyl-L-cysteine sulfoxide has also been isolated from a brassica (cabbage), and may be responsible for the production of dimethyl disulfide in cabbage. However, allinase-type enzymes are missing in brassica plants. Thus, it is not the sulfoxides, but the glucosinolates that are responsible for the biting flavour of fresh brassica and other cruciferous plants.

Glucosinolates

The isothiocyanates that give the pungent aroma to cruciferous plants appear during mechanical cell injury from the non-volatile precursors, glucosinolates. These compounds are thioglycosides, which are believed to be derived from amino acids and sugars via a common biosynthetic pathway (Figure 2.10). While some glucosinolates are clearly derived directly from amino acids, others possess side chains that require some prior degree of structural modification.

Portas (1608) was the first to publish his observations on the formation of a volatile oil from the distillation of mustard seed.[10] By the early 1800s, it was known that the pungent factor of mustard oil contained sulfur and was only formed after the mustard seeds had been pulverized in the presence of water. In 1831, a crystalline mustard oil precursor was isolated from white mustard seed (sinalbin). Soon after, in 1840, a related precursor, sinigrin, was isolated from black mustard seed. The structures of sinigrin and sinalbin were proposed in 1897 to be thioglucosides, based upon analysis of their chemical decomposition products. However, it was not until 1956 that the correct structure was published, which was confirmed in 1963 by X-ray crystallography (see the structures in Figures 2.11–2.13). Debates over these compounds were of great significance in the development of the concept of structural isomerism in organic chemistry.[7]

Myrosinase (actually a family of thioglucosidase enzymes) is responsible for cleaving off the glucose fragment to form unstable aglycones, which usually rearrange into the corresponding isothiocyanates (Figure 2.11).

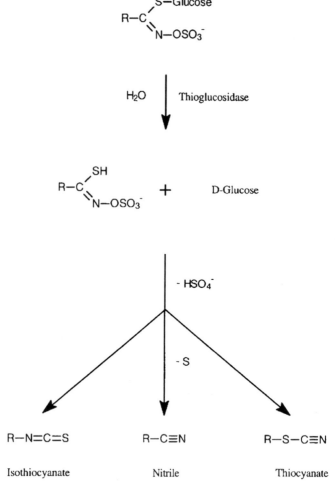

Figure 2.11 *Enzymatic production of flavour compounds from glucosinolates.*

However, depending on the structure of the glucosinolate and the presence of other compounds, rearrangements can occur to form nitriles, thiocyanates, oxazolidine-2-thiones, hydroxynitriles and epithionitriles. Under alkaline conditions, glucosinolates containing a hydroxyl group in the β-position of the side chain yield unstable isothiocyanates which undergo spontaneous cyclization to the corresponding oxazolidine-2-thione (Figure 2.12). If the pH is low (such as in the processing of sauerkraut), then the hydroxynitrile predominates. The ferrous ion also exerts a profound effect, favouring the production of hydroxynitriles. Epithionitriles are formed as a result of the

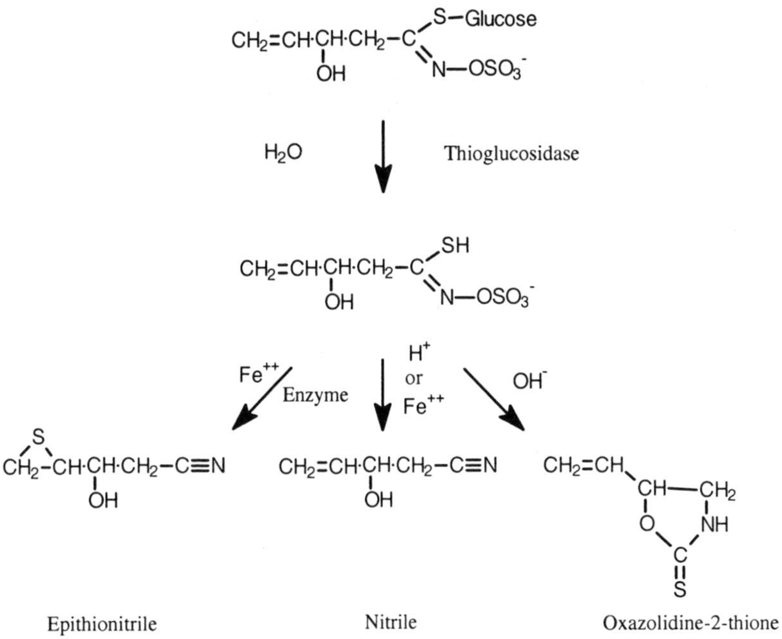

Figure 2.12 *Enzymatic production of flavour compounds from β-hydroxy glucosinolates.*

combined action of ferrous ion and a protein on myrosinase. Indole glucosinolates (Figure 2.13) yield unstable isothiocyanates which, depending upon the pH, either lose sulfur (to form the corresponding nitrile) or the entire thiosulfinate ion (to form the corresponding alcohol). Occasionally, the formation of organic thiocyanates rather than isothiocyanates following myrosinase treatment occurs, but the reasons for this are still not clear.

Terpene Reactions/Rearrangements

Glycoside cleavage. Many of the terpenoids are stored in plants as non-volatile glycosides. Thus, the essential oil of umbelliferous seeds (for example, celery seeds) is rather difficult to obtain by steam distillation. However, when a glycosidase enzyme, which cleaves the sugar off the glycoside precursor, is added to crushed seeds in water prior to steam distillation, the amount of essential oil increases dramatically.

Linalool is a minor component only of passion fruit pulp when isolated at pH 7. However, under natural conditions (pH 3.0), the concentration of linalool increases more than 50 times.[8]

Double-bond shifts, cyclizations and dehydrations. Rearrangements and dehydra-

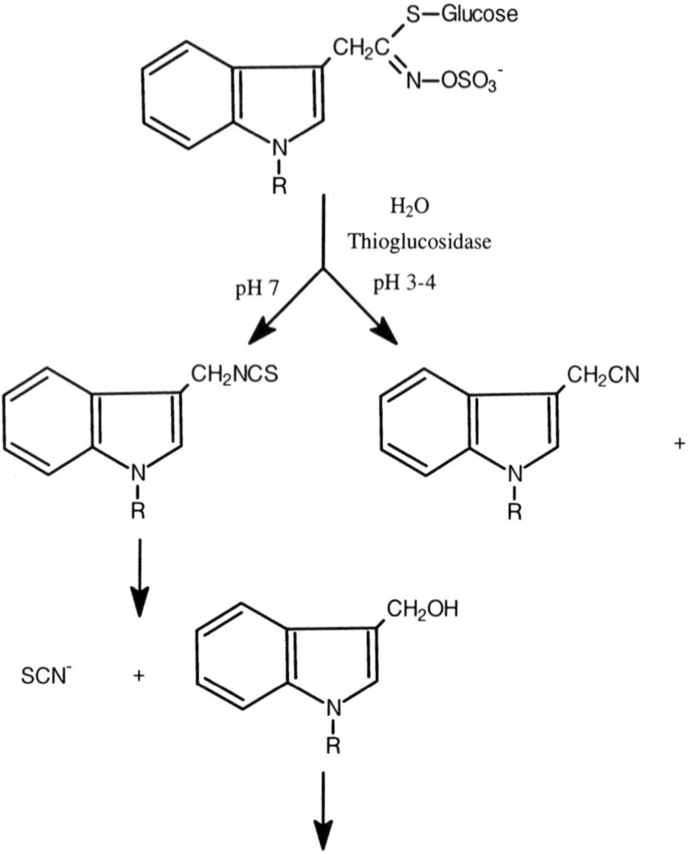

Figure 2.13 *Enzymatic production of flavour compounds from indole glycosinolates.*

tions of terpenoid compounds occur under very mild conditions. As shown in Figure 2.14, the formation of a cation will easily rearrange non-cyclic terpenes into many different bicyclic species. Only a very small amount of acid or base is needed to initiate double-bond shifts, cyclizations, and the loss of water. Thus, artefact formations are a problem during isolation as well as during the processing and storage of food products.

Whenever nerol oxide occurs with hotrienol and the diastereoisomeric deoxylinalool oxides or cyclic ethers (Figure 2.15), a situation found in teas, grapes and wines, one must suspect the diols A and B as precursors.[11] Hotrienol, which has a very sweet and flowery flavour resembling lime-tree blossoms, is not considered a direct component, but an artefact of the isolations.

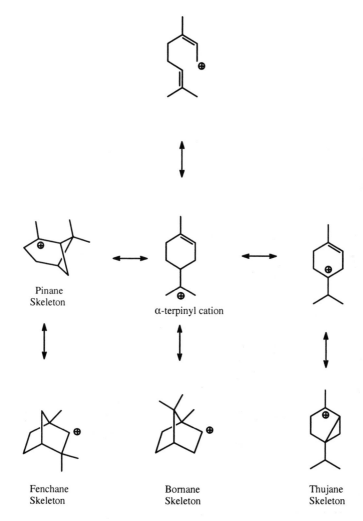

Figure 2.14 *Terpene rearrangements.*

During normal steam distillation, *p*-mentha-2,8-dien-1-ol gives more than 50% piperitenol. Only 60% of zingiberenol survives GC, giving zingiberene (16%) and β-phellandrene (20%) in addition.

Carotenoid degradations. Carotenoids are tetraterpenoid pigments which occur in all photosynthetic plant tissues as components of chromoplasts (green from chlorophyll) or chloroplasts (yellow to red from carotenoids).

Oxidation of β-carotene produces C-9 to C-13 compounds[12,13] as shown in Figure 2.16, with β-ionone having the largest aroma value due to its

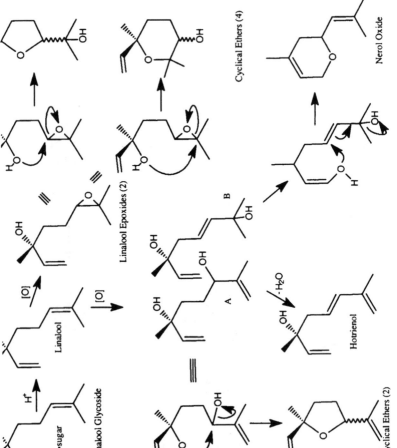

Figure 2.15   *Linalool rearrangements.*

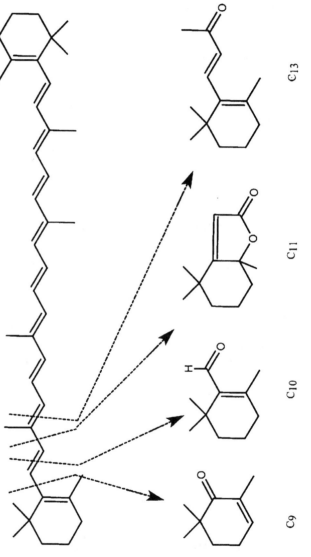

Figure 2.16 *Oxidation of β-carotene at specific double bonds yields C₉ to C₁₃ compounds. Adapted from Enzell and Wahlberg.[12,13]*

characteristic violet-like aroma and low threshold value of 0.007 ppb in water. Ionone compounds in virtually all oxidation states are found in flavours; among these are the dehydro-derivatives as well as the aromatic analogue. Used empirically in the compounding of fruit flavours, β-ionone was first discovered in 1947 in raspberries, along with α-ionone, β-ionone mono-epoxide and dihydro-β-ionone. Ionones have also been found to contribute to the flavour of tobacco, black tea, passion fruit and blackberries. The ionones have also been detected in distilled alcoholic beverages, orange oil and celery oil.

Another group of C-13 aroma compounds from carotenoids are the damascones, which naturally co-occur with the ionones. β-Damascone can be formed from the allene diol which is a photo-oxygenation product of β-ionol. It has been suggested that the allenic end group of neoxanthin is a central intermediate for the *in vivo* formation of β-damascenone. Having a threshold value of 500 ppb in water, β-damascenone occurs in rose oil, tobacco, tea, raspberry, cooked apple, grape, wine and beer.

Variations in Plant Flavours

Besides varietal differences, environmental factors, such as variations in growing temperatures, rainfall, irrigation and soil nutrients, can change the amount and type of flavour compounds present in plants. It is generally accepted that stressed plants increase production of secondary metabolites and thus produce more flavourful fruits and vegetables. However, stressed plants are also usually of lower quality when used as fresh food because they may be woody, small, or visually unappealing.

Onion and/or garlic. A comparison of Indian garlic obtained from a mountainous area (23°C average temperature) and from a climate averaging 10°C warmer showed the trend that garlic grown in a cooler climate contains less methyl and more allyl sulfur compounds. Also, refrigeration of garlic bulbs increases the levels of prop-1-enyl sulfur compounds.

Glucosinolates. The total glucosinolate content and the relative amounts of the individual glucosinolates of a particular plant depend upon the part examined as well as on the variety, cultivation conditions, climate and agronomic practices. Certain glucosinolates found in seed may not be present in the developing and mature plant.

FLAVOUR DEVELOPMENT DURING FOOD PROCESSING

Sugar Thermal Breakdown – Caramelization

When sugars are heated to 100–130°C, any bound water will be released, but without alteration to the molecular structure of the food. At temperatures between 150–180°C, a molecule of water is lost from the sugar molecule, giving an anhydride that can lose further water. This results in the formation of furfural from pentose or 5-hydroxymethylfurfural from hexose sugars. With prolonged high temperatures, caramelized flavours, including many furan derivatives, carbonyl compounds, alcohols and both aliphatic and aromatic hydrocarbons, are formed.

Caramelization. In candy manufacture, sugars may be burnt intentionally to create flavours. For example, compounds such as maltol, furaneol, nor-furaneol, isomaltol, cyclotene and maple lactones are produced during the boiling of maple sap to produce the typical flavour of maple syrup (Figure 2.17).

Amino Acid Thermal Breakdown

Enzyme activation. The flavour enhancer guanosine 5'-monophosphate (GMP) is rarely detected in raw vegetables. Rather, it is produced during the processing of vegetables from endogenous ribonucleic acid (RNA) which is hydrolysed by enzymes activated at temperatures of 65–75°C. Blanching and steaming of mushrooms (shittake), green beans and green bell peppers result in a high conversion of RNA into GMP; potatoes have an intermediate conversion and cabbage has no conversion of RNA into GMP.

Protein pyrolysis. The thermal decomposition of amino acids and peptides requires temperatures higher than those that are normally encountered during the cooking of foods. Only in surface areas of grilled foods, where localized dehydration allows the temperature to rise significantly above the boiling point of water, will decarboxylation and deamination of amino acids occur with the formation of aldehydes, hydrocarbons, nitriles and amines.

Maillard Reaction. This is one of the most important routes to flavour compounds in cooked foods. The French chemist Louis Maillard first described this reaction between reducing sugars and amino compounds in 1912 when he investigated the coloured compounds (melanoidins) formed in the heating of a solution of glucose and glycine. The Maillard reaction does not require as high a temperature as those associated with sugar caramelization and protein pyrolysis. Even mixtures of refrigerated sugars and amino acids can show signs of Maillard browning over time. However, Maillard reactions proceed much faster at higher temperatures. Both browning and the formation of flavour compounds in foods generally occur at the high temperatures associated with cooking, evaporation, heat processing and drying. Reaction rates also increase with low moisture levels. Hence, flavour compounds produced by the Maillard reaction tend to be associated with the surfaces of the foods that have been dehydrated by the heat source.

The first step of the Maillard reaction involves the addition of a carbonyl group of a reducing sugar in the open chain form with the amino group of an

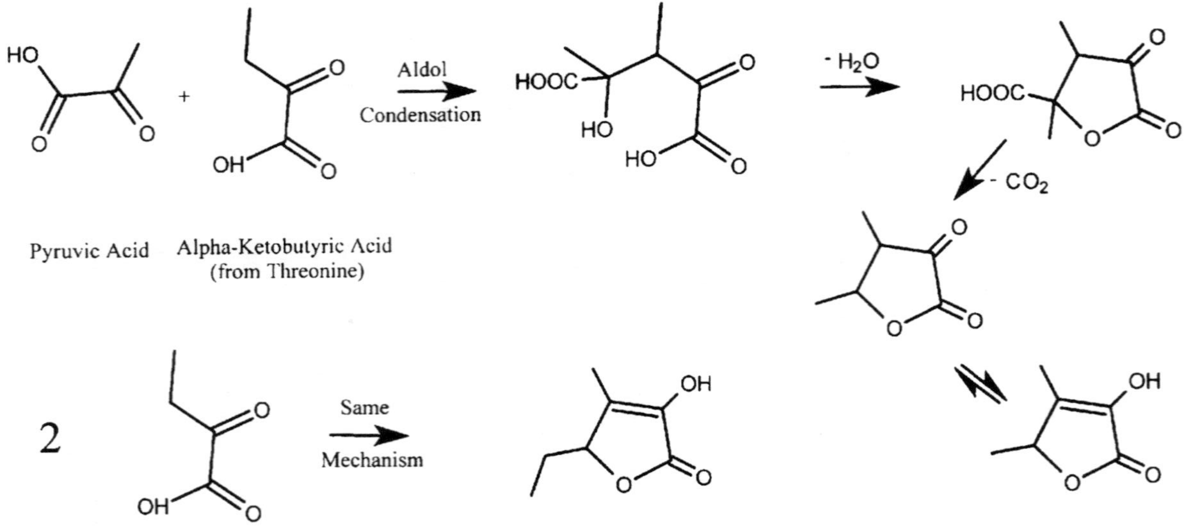

Figure 2.17 *Formation of maple lactones during the boiling of maple sap.*

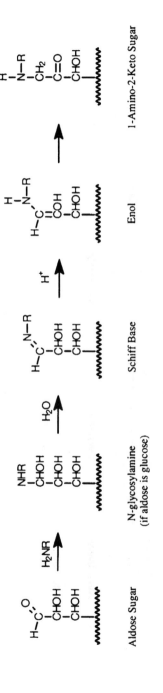

Figure 2.18 *Amadori rearrangement.*

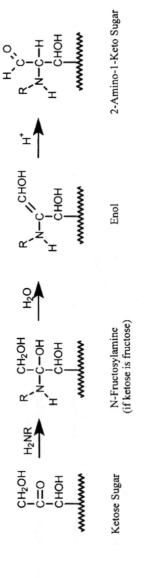

Figure 2.19 *Heyns rearrangement.*

amino acid or peptide (a primary amino group) (Figure 2.18). The subsequent elimination of water results in a Schiff base. Water is the limiting factor and, at this stage, the reaction is reversible. The Schiff's base cyclizes to give the corresponding *N*-substituted aldosylamine. This is converted into the 1-amino-1-deoxy-2-ketose by the acid-catalysed Amadori rearrangement. If a ketose, such as fructose, is involved instead of an aldose sugar, then a ketosylamine is formed. The ketosylamine undergoes the Heyns rearrangement to form a 2-amino-2-deoxyaldose (Figure 2.19). The Amadori and Heyns intermediates themselves do not contribute to flavour; however, they are important precursors of flavour compounds. They are thermally unstable and undergo dehydration and deamination (Figure 2.20) to give a host of degradation products, including furans similar to those obtained in sugar caramelization.

Strecker degradation. The Strecker degradation (Figure 2.21) is one of the most important flavour reactions associated with Maillard browning. It produces little or no browning of foods itself, because it only involves the oxidative deamination and decarboxylation of an α-amino acid in the presence of a dicarbonyl compound. The Strecker degradation then leads to the formation of an aldehyde containing one fewer carbon atom than the original amino acid and an α-aminoketone. This is an important intermediate in the formation of several classes of heterocyclic compounds, including pryazines, oxazoles and thiazoles.

The aminoketone formed from cysteine is a powerful reducing agent and probably produces hydrogen sulfide by reduction of mercaptoacetaldehyde or cysteine. Alternatively, the break down of the mercaptoimino-enol intermediate formed by the decarboxylation of the cysteine–dicarbonyl condensation product yields hydrogen sulfide, ammonia and acetaldehyde with the regeneration of the original dicarbonyl compound. As hydrogen sulfide is important for the formation of many highly odoriferous compounds in eggs and meats, this emphasizes the importance of cysteine in the development of flavours.

The Maillard reaction and associated Strecker degradation are still not fully understood. The subsequent and competing reactions produce a very complex picture, especially as products from one reaction become reactants for another. The amount and type of flavour compounds generated from the Maillard reaction vary with (1) the type of amino compounds and sugars present in foods, (2) the pH at which the reactions occur, (3) the amount of water available, (4) the presence of any salts that can buffer the pH and (5) the length of time foods are held at specific temperatures.

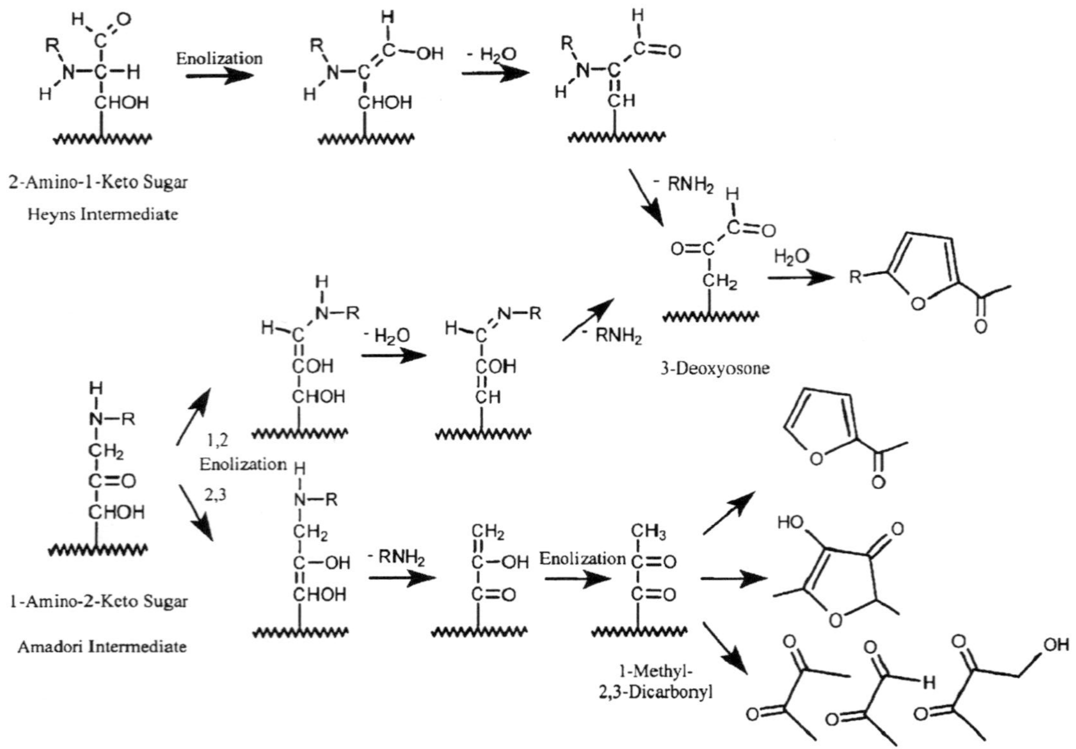

Figure 2.20 Formation of furans and dicarbonyls from Heyns and Amadori intermediates.

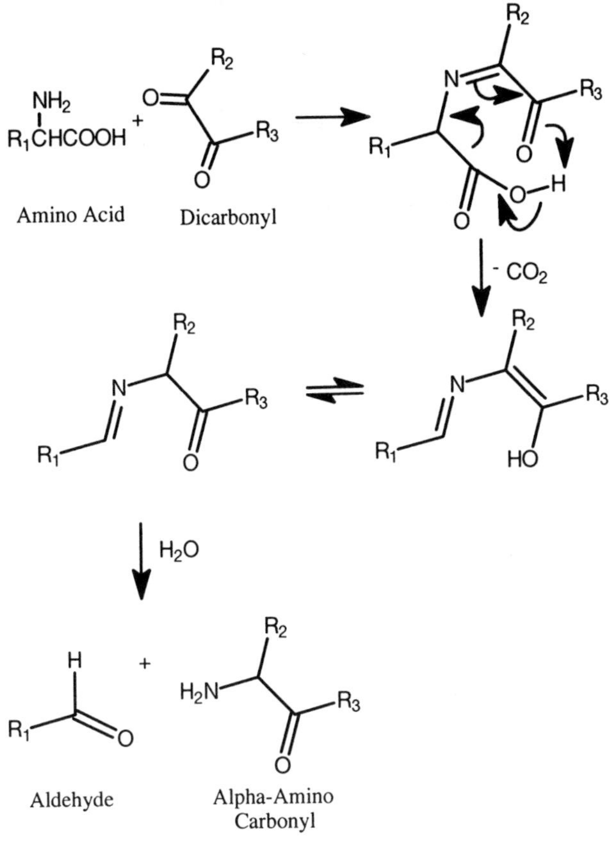

Figure 2.21 *Strecker degradation.*

Lipid Thermal Breakdown

The familiar rancid odours of deteriorated fats and oils result from autoxidation of unsaturated fatty acids. The course of the reaction is the same whether the acids are esterified or free and occurs in three steps (Figure 2.22). The *initiation* stage produces small numbers of highly reactive free radicals, molecules with unpaired electrons. In the *propagation* stage, oxygen reacts with the free radicals produced in the initiation stage to form fatty acid hydroperoxides. These hydroperoxides break down to generate more free radicals which can either attack more fatty acids and/or cycle with more oxygen to maintain a chain of reactions. When sufficiently high concentrations of free-radical species are formed, they tend to react together in *termination* reactions to give polymers or decompose to form the stable

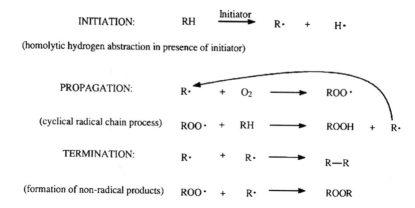

Figure 2.22 *Free radical lipid oxidation.*

end-products characteristic of rancid fat: aldehydes, alcohols and ketones. For example, *cis*-hex-3-enal, butanedione, pentane-2,3-dione, and penta-2,4-dienal give the characteristic odour of 'reverted' soybean oil.

The free radical lipid oxidation reactions involved both during cooking and in rancid flavour formation follow the same basic pathways (described above) and result in similar types of volatile products. There are differences in the precise mechanisms of oxidation under refrigeration and thermal conditions. In the case of cooking, the radical reactions contribute to desirable flavour; in a stale rancid product, the radical reactions result in off-flavours. Hydroperoxides are extremely heat labile. At lower temperatures, they are more stable, and more may build up before decomposition occurs. This leads to different proportions of the various radical intermediates, resulting in different volatile products.

The aldehydes and ketones produced by lipid oxidation can also undergo secondary reactions in the presence of amino acids to make other flavour compounds. Lipid oxidation during cooking plays an important part in the development of the complex profile of meat aroma volatiles; in lightly grilled or boiled meats, especially poultry, such volatiles have been found to dominate.

However, lipid autoxidation also occurs during the long-term storage of meats, producing undesirable off-flavours. Warmed-over flavour, a flavour that becomes apparent when cooked meat is kept at refrigerator temperatures for a few days and is then reheated, has been attributed to lipid oxidation products. It is believed to be due to the autoxidation of phospholipids, which are usually more unsaturated than triglycerides. During the initial heating, disruption and dehydration of meat cell membranes, the phospholipids become more susceptible to oxidation. This reaction is catalysed by traces of metal ions, in particular iron, made available by the break down of haem pigments during cooking.

Thermal Lignin Degradation

The smoke processing of fish and meats has a long tradition in the preservation of foods. However, the process also adds unique flavour properties. Today, the smoking of foods is used more to add flavour than for preservation.

Although phenols, as a compound class, are considered the major contributors to wood-smoke aroma, they are not totally responsible. The major components of wood, hemicellulose, cellulose and lignin, thermally degrade to produce flavour volatiles. Hemicellulose degradation produces lactones and furans, while cellulose degradation produces anhydro-sugars that can be involved in further classic carbohydrate thermal degradations.

The major phenols found in wood smoke, guaiacol (*o*-methoxyphenol), 4-methylguaiacol, syringol and phenol, are directly related to lignin degradation. Initially, fission occurs in the heterocyclic furan/pyran rings and ether linkages of lignin, to produce ferulic acid which degrades to give 4-vinylguaiacol. This in turn degrades to form vanillin and guaiacol, which further degrade to phenol and cresols. In hardwoods containing higher amounts of methoxy substitution, dimethoxy rings result in the formation of syringol and para-substituted phenols.

Thermal Reactions of Sulfur-Compounds

Cooked onion/garlic aromas. The methods by which onions and garlics are cooked result in various flavour profiles. If there is a lot of water present during heating, many di-, tri-, and poly-sulfides are formed from thiosulfinates (accumulated during enzyme release when onions and garlics are cut; see p. 33). If, on the other hand, very little water is present, and the alliums are in lipids (such as in oil-macerated garlic or when onions are stir-fried in vegetable oil), many more complex poly-sulfur compounds are formed, including vinyldithiins and ajoenes (Figure 2.23).

THE USE OF BIOTECHNOLOGY TO DEVELOP FLAVOURS

Microbes

Fermentation has been practised for the production of food since ancient times. Yeasts have been utilized in brewing alcoholic beverages and in bread making. The major product of yeast metabolism is ethanol, but it also

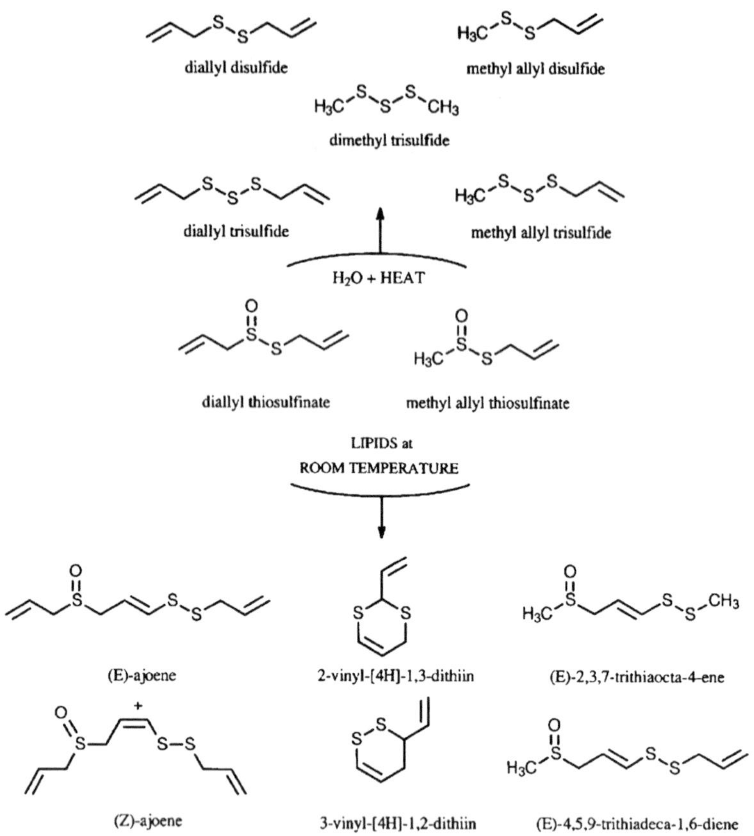

Figure 2.23 *Flavour development of cooked garlic varies with temperature and medium.*

produces simple aliphatic and aromatic alcohols, fatty acid esters, carbonyls, lactones, thio-compounds and some phenolics. The ability of many yeast strains to form high concentrations of lactones has found industrial application. In yeast strains where β-oxidation results in reduced 3-hydroxylation of fatty acids and increased production of 4- and 5-hydroxylations, the cytotoxic effect of free fatty acids in the fermentation broth is reduced and high concentrations of triglycerides in the fermentation broth stimulate the formation of lactones. To increase yields, lactone bioreactors constantly remove volatile flavour products from the fermentation broth as they are produced.

During the ripening process of cheese, a number of enzymatic reactions are catalysed both by microbial enzymes and enzymes already present in raw milk. Milk fat, protein and carbohydrates are degraded to varying extents. These all give rise to a very complex mixture of compounds, some of which contribute to characteristic cheese flavours.

Plants. Through the use of plant tissue culture techniques, especially combined with normal genetic breeding techniques, cell lines of food plants (fruits, vegetables, spices and herbs) can be improved. The plant tissue techniques include (1) micropropagation of plantlets, (2) regeneration of plants from callus, (3) protoplast fusion to produce new hybrids and (4) gene transfer.

Although used as much as a colourant than as a flavourant, saffron is one of the most expensive food additives used today. It is very slow-growing, has specific climatic restrictions, and has so few stamens per flower. This plant is a good example of how plant tissue culture is being used because normal breeding is inefficient.

Much research has also been conducted to develop *in vitro* production of flavour compounds using suspended plant cell cultures. However, commercialization of this technique has been hindered by the fact that it is usually cheaper simply to extract the intact plant for the flavour chemicals. There are several problems associated with the large scale production of flavour compounds *in vitro:*

1. A superior high-yielding cell line is necessary, thus requiring considerable screening of parent lines and the development of analytical techniques for measuring yields.
2. The biochemical pathways of flavour production are not fully understood.
3. The synthesis and/or storage of flavour compounds may be in specialized cells or organs.
4. Culture medium is often expensive; also, because plants produce flavours as secondary metabolites, two media are usually needed.
5. Cytotoxic flavour compounds, which make the culture unstable, must be removed frequently to avoid low yields.
6. New equipment needs to be designed to address problems associated with plant cell shear sensitivity, oxygen transfer, and cell aggregation.

Even with all these challenges, plant cell cultures have potential application in the marketplace. Industrial production of shikonin, a naphthoquinone, by a two-stage cell culture process by the Japanese was necessary (and successful) because domestication of the wild shikonin plant, *Lithospermum erthrorhizon,* failed.

Enzymes

In plants and microbes, the only bioconverters are enzymes. Enzymes, either in a crude or a purified form, have traditionally been used as processing aids in the food industry. Enzymes have a high catalytic activity towards a specific substrate and produce end-products under mild controllable conditions. But they do have limitations: (1) they are not stable under extreme conditions, (2) they have limited substrates and (3) water must be present for their activity.

One of the very first *in vivo* usages of enzymes to produce flavours on a large scale is a product known as lipolysed milk fat. This product utilizes lipases to liberate fatty acids and has found wide application for the enhancement of butter and cheese characters in dairy products, and for flavour development in milk chocolate.

Enzyme modified cheeses (EMC) are products obtained by a controlled proteolytic and/or lipolytic enzyme treatment of a previously manufactured traditional natural cheese. The most important flavour reactions are the liberation of volatile fatty acids from milk fat and the degradation of casein into low molecular weight peptides and amino acids without bitterness development. After thermal inactivation of the enzymes, the EMC product obtained can easily have a flavour intensity 20 times greater than that of natural cheese. EMC is usually used internally in processed cheese manufacturing.

In assessing the potential uses of enzymes for the production of flavour compounds, one should consider that in most plant tissues many flavour constituents are bound as glycosides. There is increasing industrial research activity being devoted to the isolation and separation of specific glycosidases from suitable sources. These should soon be available as processing aids to enhance the flavours inherent in many vegetables and fruits.

CONCLUSION

As is illustrated in this chapter, flavour compounds range through the spectrum of chemical compounds and can be both volatile and non-volatile. Flavour compounds can sometimes occur naturally in raw foods, but often they need to be developed either enzymatically via cell-wall cleavage or thermally during cooking. As we learn more about flavours and how they are produced, better-tasting foods are becoming more readily available at the marketplace.

The reality of good food production is in learning how to control flavours during manufacturing. Processing conditions can be maximized for flavour constituents and the resulting isolates which can be utilized to control the flavour of foods. But in order to control flavour, one must be able to analyse the flavour components both qualitatively and quantitatively.

REFERENCES

1. G. Ohloff, in *Progress in the Chemistry of Organic Natural Products*, eds. W. Herz, H. Grisebach and G.W. Kirby, Springer-Verlag, New York, 1978, Vol. 35, p. 431.
2. F. Rijkens and H. Boelens, *Proceedings of the International Symposium Aroma Research*, Zeist, Pudoc, Wageningen, 1975, p. 203.
3. C. Weurman, S. van Straten and F. de Vrijer, *Lists of Volatile Compounds in Food*, Central Institute for Nutrition and Food Research, Zeist, Holland, 3rd edn., 1973, with suppl. 1 to 7, 1976.
4. M.E. Knowles, J. Gilbert and D.J. McWeeny, *J. Sci. Food Agric.*, 1975, **26**, 189; A.O. Lustre and P. Issenberg, *J. Agric. Food Chem.*, 1970, **18**, 1056.
5. K. Torii and R.H. Cagan, *Biochim. Biophys. Acta*, 1980, **627**, 313.
6. J.I. Suzuki, F. Tausig and R.E. Morse, *Food Technol.*, 1957, **11**, 100.
7. T. Suzuki and K. Iwai, in *The Alkaloids*, ed. A. Brossi, Academic Press, New York, 1984, Vol. 23, p. 227.
8. L.D. Lawson and B.G. Hughes, *Planta Med.*, 1992, **58**, 345.
9. E. Block, S. Naganathan, D. Putman and S.-H. Zhao, *J. Agric. Food Chem.*, 1992, **40**, 2418; E. Block, *J. Agric. Food Chem.*, 1993, **41**, 692.
10. A.J. MacLeod, *Flavour Ind.*, 1970, **1**, 665.
11. G. Ohloff, I. Flament and W. Pickenhagen, *Foods Rev. Int.*, 1985, **1**, 99.
12. C.R. Enzell and I. Wahlberg, *Recent Adv. Tob. Sci.*, 1980, **6**, 64.
13. C.R. Enzell, in 'Flavour '81', 3rd Weurman Symposium, Munich 1981, ed. P. Schreier, Walter De Gruyter, Berlin, 1981, p. 449.

Chapter 3

The Chemical Senses

INTRODUCTION

The chemical senses are the most primitive of the specialized sensory systems, with an evolutionary history of some 500 million years. It is fitting, then, that they deal with the most basic biological requirements: feeding, to preserve the organism, and reproduction, to preserve the species. While humans rely heavily on vision and hearing, our primary sensory legacy is taste and smell. We are unusual animals in this regard because of our recent ecological niches, aloft in trees and lifted up on two feet. Our sensory apparatus is clear of the taste- and odour-rich earth, mounted sufficiently high to offer a vantage that gives value to the straight lines demanded by vision. But chemical senses dominate the animal kingdom, and so should be expected to have pervasive, if subtle, effects on human evolution.

The sense of taste largely manages dietary selection, not only by its analysis of the quality and concentration of potential foods, but also through communication with the gut. Smell is used to identify predator and parent, and to find food, mate and home. Both chemical senses are intricately woven with two neural systems: those serving emotions and memories.

One rarely has a gustatory or olfactory experience without an accompanying emotional reaction. While we may use eyes and ears to survey our environment continuously and impassively, the chemical senses are engaged only occasionally, and always with passion. Odours and tastes carry emotional components of like or disgust, motivating the approach or withdrawal that determined which of our predecessors would survive to become our ancestors. We have inherited that hard-won and deeply ingrained knowledge. We call it 'body wisdom'.

The chemical senses also generate extraordinarily acute and enduring memories. With taste, these have to do with the internal consequences of ingestion. Passionate preferences are created for those tastes that accompany

56

nutrition. The Italian child with garlic on his pasta, the Japanese with soy beside her rice and the Mexican with pepper on his tortilla, each develops a preference for the taste identified with the accompanying carbohydrate load. From these preferences emerge the distinct cuisines that help characterize diverse human cultures.

The taste of Proust's madeleine transported him to a happy childhood, but not all gustatory memories are so kind. When a distinct and novel taste is followed even hours later by nausea, a powerful aversion is created and that taste is assiduously rejected thereafter. These learned aversions emerge from a single experience and may last a lifetime.[1]

Odour perceptions are well known to evoke childhood memories, even in visually oriented humans. In lower animals, the memorial feats associated with smell are extraordinary. This might be expected from the anatomical structure of the olfactory system, which branches and reaches several brain areas thought to be involved in memory formation and maintenance.

Thus, smell and taste are the senses that have most shaped our evolutionary history. They guide our most primitive biological functions by evoking intense emotions of pleasure or revulsion that are stored for a lifetime.

In this chapter we analyse these chemical senses, with particular reference to the perception of flavour they offer, and to the selection of foods that they guide.

ANATOMY OF THE CHEMICAL SENSES

Despite the fact that olfaction and taste are closely linked in the appreciation of flavour, their anatomical structures are quite independent. Olfaction enters the brain from the front, sending fibres to the olfactory cortex after only one synapse in the olfactory bulb. Taste approaches from the rear, entering the brain in the medulla, and working its way through two synapses towards the gustatory cortex. Only after completing these separate courses do olfactory and gustatory fibres project to common areas – and indeed common neurons – in secondary cortex and in subcortical areas such as the amygdala and hypothalamus.

Olfaction

The olfactory system is composed of three anatomically discrete regions, the *olfactory epithelium*, the *olfactory bulb* and the *olfactory cortex*. The epithelium is the receptor surface where odorant molecules are bound. Cells here send their axons directly to the olfactory bulb, which extends forwards from the main mass of the brain to accept them. After extensive local processing, the two

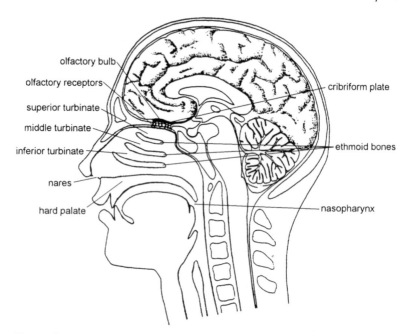

Figure 3.1   *Lateral view of the human head showing the location of major structures in the nasal area.*[2]

projection neurons of the bulb – the mitral and tufted cells – send olfactory information to the olfactory cortex, from which it is widely distributed throughout the nervous system.

Olfactory epithelium. This sheet of olfactory receptor tissue is located along the top of the nasal cavity, in the septum and in parts of the superior turbinates (Figure 3.1). Moving down along the turbinates, there is a gradual transition from the graceful cilia of olfactory cells to the stout projections in the respiratory epithelium.

Olfactory epithelium comprises three main cell types: receptor cells, supporting cells and basal cells (Figure 3.2). Receptor neurons are small (5 µm) and ovoid. They are created at the base of the epithelium and send long, slender dendrites up through its 30 µm thickness to emerge at the surface. As these cells proceed through their two-month life cycles, they migrate towards the surface, and the apical dendrites shorten and thicken. Where the dendrite meets the surface, it forms a knob from which many cilia emerge to spread 30 µm and more across the surface of the receptor tissue. Their surfaces are rich with particles that are the presumed sites for binding odorant molecules.[3]

From the basal side of the receptor cell, a fragile, non-myelinated axon – at 0.2 µm among the most slender in the nervous system – proceeds to the

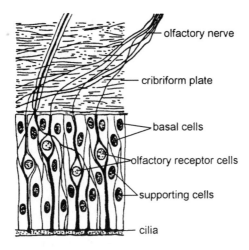

olfactory nerve

cribriform plate

basal cells

olfactory receptor cells

supporting cells

cilia

Figure 3.2 *Cross-section of the olfactory epithelium, showing basal, supporting and olfactory receptor cells and their central projections through the cribriform plate.*[2]

olfactory bulb. As they leave the epithelium, the axons are packed into bundles of up to 200 by Schwann's cells. The bundles then join to form small nerves that penetrate the *cribriform plate* of the ethmoid bone before spreading across the surface of the olfactory bulb (Figure 3.2). Passage through the plate is dangerous. There is no cushion of fluid or tough membrane to protect the delicate fibres, which are thus subject to damage when the head suffers trauma. Loss of smell is a common result of head injury, and often results from the severing of olfactory axons at the cribriform plate. Fortunately, the sense of smell usually recovers as new receptor cells are created at the base of the epithelium, and send their axons on to the bulb.

The number of olfactory receptor cells is huge, particularly in *macrosmatic* animals, such as rabbits where they approach 100 million in number. In humans and other *microsmatic* primates, the entire olfactory system is relatively small, having been displaced by the enlarged cerebrum, under whose bottom surface it is tucked. The adult human possesses merely 10 million receptor cells. Still, only the rods and cones of the eye outnumber olfactory receptors among the sensory neurons.

Supporting, or *sustentacular*, cells form an orderly layer in the upper one-third of the epithelium. They have rectangular cell bodies, and terminate in *microvilli* at the epithelial surface. However, they have no axons and are not thought to be involved in carrying sensory information.

Basal cells are of 6 μm diameter and are located at the bottom of the epithelium. They are the progenitors of receptors, constantly replacing the

200 000 receptor neurons that die each day. The signals that cause them to divide and differentiate are not known.

The continuing loss and re-establishment of connections between the epithelium and the bulb raises questions of how organization can be maintained. It is clear that there is not a simple mapping of the epithelium onto the surface of the bulb, as there is, for example, between the retina and the lateral geniculate body in vision. Rather, projections from specific regions of the epithelium go densely to focal sites in the bulb, but then diffusely to areas around that site. Conversely, neighbouring epithelial cells may project to quite different areas of the bulb. It is thought that individual axons, which may originate from diffuse regions of the epithelium, are segregated into bundles that carry similar information and sent to a particular location in the bulb. Therefore, the crude topographic relationship between epithelium and bulb is overlaid by one of functional specificity. Each receptor cell that is created must carry a functional tag that determines the course of its axon to the bulb.

Olfactory bulb. The olfactory bulb is a forward projection of the brain, divisible into six layers among whose cells there is a complex interplay of excitation and inhibition (Figure 3.3). From the surface to the core, the layers are (1) the *olfactory nerve layer*, composed of incoming fibres from the olfactory epithelium, (2) the *glomerular layer*, where the main interactions between first- and second-order fibres occur, (3) the *external plexiform layer*, characterized by a plexus of dendrites rising from beneath, (4) the thin *mitral cell layer*, composed of mitral cell bodies, the bulb's most prominent cells, (5) the *internal plexiform layer*, a cell-poor region composed of the axons and axon collaterals of mitral and tufted cells and (6) the *granule cell layer*, comprising neurons that mediate interactions among the mitral cells.

Distributed among these layers are five classes of neurons, two of which receive information from the epithelium and project it to the cortex, and the others to mediate local interactions.

The bulb's largest and most distinctive neuron is the *mitral cell*, shaped like a bishop's mitre in frontal profile. The adult human possesses about 50 000. The 30 μm soma of this cell has three projections: a prominent, 2–12 μm diameter, apical dendrite that rises radially a distance of 600 μm, up through the external plexiform layer to invade and ramify throughout a glomerulus; a complex of 2–9 secondary dendrites that distribute themselves tangentially along the external plexiform layer and extend for hundreds of microns, even to the contralateral half of the bulb; and a myelinated axon, 0.5–3.0 μm in diameter, that emerges from the basal point of the mitred soma to join others in forming the lateral olfactory tract (Figure 3.3).

The second projection neuron of the bulb is the *tufted cell*, with its characteristic crest from which it derives its name. The bodies of tufted cells

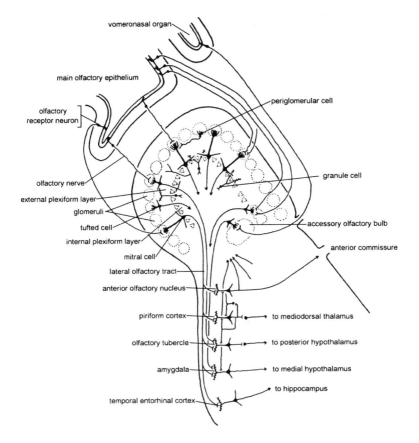

Figure 3.3 *A schematic representation of the olfactory system, with central focus on the olfactory bulb. The bulb receives input from the olfactory epithelium by way of the olfactory nerve, processes that input and project to five areas of the cortex by way of the lateral olfactory tract. Adapted from Shepherd and Greer.*[4]

are scattered throughout the considerable breadth of the external plexiform layer, and typically measure 20 μm. Their three projections are similar to those of mitral cells: an apical dendrite that rises to a glomerulus, within which it arborizes extensively to contact axons coming from the epithelium; secondary dendrites that ramify through the superficial part of the external plexiform layer; and a myelinated axon that joins with those of mitral cells to form the lateral olfactory tract (Figure 3.3).

Interactions within glomeruli are complex; they are orchestrated in part by local neurons, *periglomerular cells*, whose 8 μm somata lie at the base of the glomeruli. Each of these cells sends a small dendritic plexus into its associated glomerulus, and a short axon that projects tangentially to form synapses within a radius of five glomeruli. Thus, periglomerular cells are the first example of intrinsic interneurons that function completely within the bulb (Figure 3.3).

The second class of interneuron is the *granule cell*, whose bodies compose the sixth layer of the bulb. These are the most numerous cells in the olfactory bulb, numbering about 2 million in the rat, and, as their name implies, they are small – typically 6 μm in diameter. The granule cell body sends out a small tuft of dendrites in its immediate vicinity to the basal side, but its main projection is a slender apical dendrite that rises through the mitral cells and spreads widely throughout the external plexiform layer (Figure 3.3). There is no axon. The cell has a morphology that suits it to mediate local interactions, and so, both in shape and function, the granule cell is analogous to the amacrine cell of the retina (Figure 3.3).

Finally, there is a miscellaneous collection of *short axon cells* distributed throughout the external plexiform and granule cell layers. These have oval cell bodies with diameters of 12–16 μm, dendrites that ramify between the glomeruli and axons that extend across a distance of up to three glomeruli. Little is known of their connections or functional significance.

Two main processes occur in the olfactory bulb: the interactions between the peripheral nerve and the mitral and tufted cells, whose axons convey the results to the cortex, and the modulation of that interaction through a complex of local circuits. Peripheral nerve axons branch widely upon entering a glomerulus and make excitatory synapses on the dendrites of mitral, tufted and periglomerular cells. In addition, the dendrites of mitral and tufted cells form reciprocal, excitatory synapses with those of periglomerular cells. Thus any axon entering the bulb from the olfactory epithelium may initiate a cascade of excitation that engulfs a local region of the glomerulus. Now the inhibitory processes come into play. The axons of periglomerular cells extend across glomeruli, forming inhibitory synapses with the apical dendrites of mitral and tufted cells. This system of *lateral inhibition* serves to sharpen the distinctions among incoming signals because, as one glomerulus rises in excitation, its activity dampens that around it, creating a bolder relief.[5]

The second site of local interaction is in the external plexiform layer, where the rising apical dendrites of granule cells form reciprocal synapses with the secondary dendrites of mitral and tufted cells. The complexity of these interactions is increased by two additional influences. First, collaterals branch from the lateral olfactory tract – composed of the axons of mitral and tufted cells – and return to the bulb to form synapses on the small sprays of dendrites that sprout from the basal side of granule cells (Figure 3.3). Thus, the cells that influence the activity of output neurons are themselves impacted by that activity. Secondly, neurons from several sites in the hindbrain and forebrain send centrifugal axons to the bulb, where they form connections at all three levels of synaptic interaction: the glomeruli, the external plexiform layer and the granule cell layer. These are excitatory synapses, which occur predominantly on the two primary interneurons of the bulb, the periglomerular and granule cells. Thus, the signal that is projected through any axon of a mitral or tufted cell is a product of at least the following: the transduction processes in the epithelium (through the olfactory nerve), the activity in glomeruli that surround the one from which the axon emerges (through periglomerular cells), its own activity from the previous instant (through collaterals to granule cells) and the condition of the individual (through centrifugal axons at all levels).

Central connections. The axons of projection neurons in the bulb – mitral and tufted cells – form the *lateral olfactory tract*. This is the major pathway to the central targets of olfaction. It should be noted, however, that most mammals also possess a robust *accessory olfactory bulb*, thought to be specialized for processing *pheromones*. The output neurons of this structure form the medial olfactory tract that projects in parallel with the lateral tract (Figure 3.3). The accessory bulb and medial olfactory tract recede to only minor status in humans, and so are not considered above. However, there are indications that humans may be responsive to olfactory stimuli that bear on reproduction, if not on feeding, through the accessory system.

Olfaction serves a variety of functions: analysis of foods, sexual arousal, emotional responses of aggression and fear, recognition of belonging (conspecifics, siblings, offspring), recognition of safety and danger and memory, particularly for events rife with emotion. Its central connections reflect this variety. The major targets of the olfactory tracts are: (1) the *anterior olfactory nucleus*, and thence through the anterior commissure to the contralateral olfactory bulb; (2) *piriform cortex*, which is the major cortical relay for olfaction, from which axons project to the mediodorsal thalamus, and then on to the orbitofrontal cortex where olfactory, gustatory and visual inputs converge; (3) the *olfactory tubercle*, from which projections continue to the posterior hypothalamus, perhaps to mediate odour-induced aggression; (4) the *amygdala*, then to the medial hypothalamus, perhaps to mediate fear responses to odours; (5) the *temporal entorhinal cortex*, and then to the hippocampus to provide the extraordinary memories associated with smell (Figure 3.3).

The olfactory cortex is primitive. Indeed, it was probably the need to develop a more sophisticated sense of smell that drove the evolution of cortical tissue, first in crocodilians, then in mammals. That pioneering region was pressed relentlessly to the side as neocortical areas continued to emerge to serve other functions, such that it now resides on the underside of the temporal lobe. The early origin of the olfactory cortex is also reflected in its structure. It has only three layers (most of the cerebral cortex has six), and is dominated by the *pyramidal cells* of layers 2 and 3. These send apical dendrites into the superficial layer 1 to receive input from the olfactory bulb. Throughout all three layers are short-axon association cells that influence, in unknown ways, the interplay between axons of the lateral olfactory tract and apical dendrites of pyramidal neurons. Further complexity is introduced by collaterals from the exiting pyramidal cell axons. These double back to terminate both on the short axons cells and on other pyramidal neurons, even at some distance from their own. Thus, when the lateral olfactory tract is stimulated with a synchronous volley, the net effect in the olfactory cortex is an initial wave of excitation followed by a prolonged period of inhibition, presumably mediated by association neurons and feedback loops.[6]

Taste

Olfactory projections approach the brain from the front, and in isolation. The gustatory system arrives from the rear, in conjunction with touch and *visceral* senses.

Gustatory receptors. Taste receptor cells are confined to *taste buds*, which are goblet-shaped structures mostly contained within small mounds on the tongue called *papillae*. In mammals, taste receptors are located exclusively in the mouth, although in other animals they reside on special structures, such as antennae or barbels, or cover the entire surface of the body.

Among the 50–150 cells that occupy the taste bud, there are four types: *basal cells, dark* (type I) *cells, intermediate cells,* and *light* (type II) *cells.* Basal cells are small and round; they are found at the bottom of the taste bud, and are thought to be the stem cells from which receptor neurons are born.

The other three cell types are elongated and bipolar, and all are believed to be receptor cells at different stages of development. Each extends from the base of the bud to the surface of the tongue, and sends fine microvilli out through the *taste pore* to sample the environment. The basal cells are thought to differentiate into dark cells, which mature into intermediate cells, and grow to senescence as light cells over a period of some two weeks. All three types receive their nerve supplies from the afferent fibres of one of the four cranial nerves that serve gustation.

Most taste buds are located on the tongue, although some exist on the soft palate, larynx, pharynx and epiglottis. Lingual buds are clustered in papillae, while those beyond the tongue are collected in sections of the epithelium. Papillae are the most obvious structural feature of the taste system; they are of three anatomical types (Figure 3.4). *Fungiform* (mushroom-shaped) *papillae* cover the anterior two-thirds of the tongue, numbering about 200 on

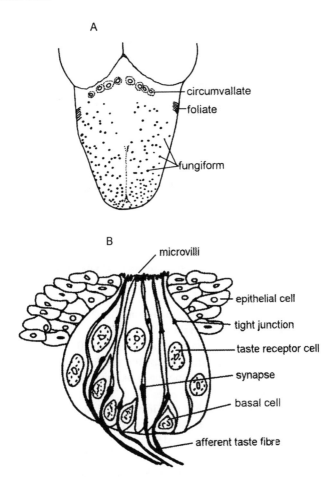

Figure 3.4   *A: Distribution of the three types of papillae that contain taste buds across the human tongue. B: Cross-section of a human taste bud.*

average, although with great variability among people. Each may contain 0–36 taste buds, with an average of about three. Thus, the typical human has about 600 buds residing in fungiform papillae. *Foliate* (leaf-shaped) *papillae* exist along the lateral edges of the tongue as five or six vertical folds in the epithelial surface. Each papilla contains an average of about 100 taste buds, and the mean total for the human tongue is approximately 1200 (6 papillae per side ×2 sides ×100 buds per papilla). *Circumvallate* (surrounded by a trench) *papillae* exist as 7–11 prominent swellings at the back of the tongue, arranged to form a chevron. The trenches that define the outer limits of each

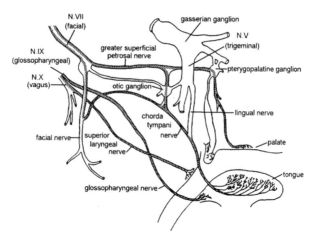

Figure 3.5 *Diagram of peripheral taste pathways (shaded lines) from the anterior and posterior tongue, the soft palate, pharynx, epiglottis and upper oesophagus. Adapted from Schwartz and Weddell.*[8]

papilla are rife with taste buds, an average of 250 in each for a total population of about 2200 buds in the circumvallate papillae. Taken together, the number of buds on the tongue averages some 4000. Another 2000 or so exist on the soft palate, pharynx, larynx and epiglottis. Taste sensitivity in humans has been shown to be a function of bud density, and this varies by a factor of 100 across individuals.[7]

The surface of the tongue has a fourth, ubiquitous type of papilla, the filiform (thread-like). These are coarse structures that give the tongue its rough texture and are not associated with taste.

Peripheral nerves. Human gustatory anatomy has not been defined in detail. The following description comes from studies of Old World monkeys, whose sensory systems are very similar to those of humans.

The peripheral nerves of gustation present a stark contrast with the anatomical simplicity of olfaction (Figure 3.5). They number four, and mix taste information with touch, temperature and pain sensations on the sensory side, and with motor fibres serving both somatic and autonomic functions. This degree of complexity led to long-standing controversies that were not resolved until the 1930s.

Fungiform papillae from the anterior two-thirds of the tongue and the more forward of the foliate papillae are served by the *chorda tympani nerve*, a branch of the *facial* (seventh) *nerve*. The fibres leave the tongue as part of the *trigeminal* (fifth) *nerve*, hence one source of controversy over their path. However, they soon separate to traverse the tympanic bulla, from which the nerve derives its name, and join the *greater superficial petrosal nerve*, which

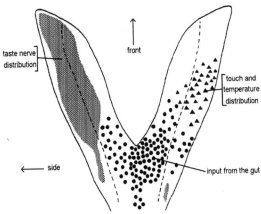

Figure 3.6   *Horizontal section through the nucleus of the solitary tract indicating the distribution of taste and non-gustatory inputs. Adapted from Hamilton and Norgren.*[9]

innervates the palate, to form the intermediate nerve, the sensory component of nerve VII.

From the remaining posterior foliate papillae, the circumvallate papillae, the pharynx, and the palatoglossal arches, both taste and somesthetic information pass through the lingual tonsillar branch of the *glossopharyngeal* (ninth) *nerve*. The superior laryngeal branch of the *vagus* (tenth) *nerve* serves the extreme posterior tongue, the oesophagus and the epiglottis.

Medulla. In contrast to the confusion that surrounded the course of peripheral taste nerves, the location of the first central relay in the medulla was established during the nineteenth century. It is the *nucleus of the solitary tract* (*NST*), a structure shaped like rabbit ears, and so-named because it is the only nucleus that lies in its particular longitudinal plane running along the brain stem (Figure 3.6). The organization of NST is determined by that of the peripheral nerves. The facial nerve enters the extreme front of the NST, where taste fibres fill the full width of the nucleus. The glossopharyngeal nerve fills an area that overlaps part of the facial nerve distribution, then extends further back. The vagus terminates furthest to the rear among the taste nerves in NST.

The front half of NST serves as the relay nucleus for taste information, while input from the gut fills the back part of the nucleus. Among the information that reaches the back parts are stretch, touch and temperature originating in the mouth, oesophagus, liver, intestines, heart, lungs, aorta, carotid body and sinus.[10] It is entirely proper that taste, which determines which materials will enter the body, should be in close communication with visceral input that reports the consequences of digesting that material.

It is estimated that only 20% of NST taste cells project to higher areas of the brain. The remainder either form connections within the NST or contribute to pathways that serve somatic and digestive reflexes associated with eating. Clusters of neurons in NST control the reflexes for acceptance or rejection of food (*cf.* section on The Control of Eating below). These reflexes are stereotypical for each of the basic taste qualities and are unaltered by the loss of all brain tissue above the brain stem, *i.e.*, they are fixed and primitive. They serve two functions. First, they allow the person to deal with the chemical in the mouth, to swallow if appealing and reject if the opposite. Secondly, they communicate a *hedonic* dimension to others. If we strip someone of the inhibitions imposed by table manners, watching his or her face will reveal a great deal about the taste experience. Thus, the first role of taste is to control reflexively the acceptance or rejection of chemicals.

A second projection from NST influences digestion. There are several pancreatic and gastrointestinal reflexes that are initiated by taste, then carried on by the stomach and intestines as they accept the nutrient load. The best understood is the cephalic phase insulin response. Thousands of β-cells in the pancreatic islets of Langerhans are stimulated to release insulin by a small number of axons coursing through the gastric and hepatic branches of the vagus nerve. These fibres originate in the dorsal motor nucleus of the vagus which, in turn, is overlaid by the gustatory NST, as a cloth might lie on an arm. Neurons in the dorsal motor nucleus send apical dendrites into the NST, effectively fusing the two structures, and offering taste input a direct influence over this autonomic reflex.

Thalamus. Other fibres project forwards to the medial ventrobasal complex of the thalamus.[11] The locus of termination has the ungainly title *ventroposterior medial nucleus pars parvocellularis* (VPMpc) (Figure 3.7). Taste cells comprise about one-third of the neurons in VPMpc. Others respond to touch or temperature signals from the mouth, or to the anticipation of an approaching taste stimulus.

Gustatory cortex. From VPMpc, taste axons project to the *anterior insula* and *frontal operculum* of the cortex (Figure 3.7). The involvement of the insular–opercular cortex in taste reception is well documented in both clinical and experimental studies. A dozen patients who suffered bullet wounds in this area had their taste disrupted. Even within this gustatory cortex, however, taste cells comprise only about 5% of the neurons. Others are associated with functions, some of which are familiar from the VPMpc: mouth movements, touch and temperature in the mouth, extension of the tongue and anticipation of an approaching taste stimulus.

If the insular–opercular region may be considered as the primary gustatory cortex, then its target, the *caudolateral orbitofrontal cortex (OFC)*, is the second-order cortical taste area. Here connections become complex. Not only is there is a direct pathway from insular–opercular to orbitofrontal

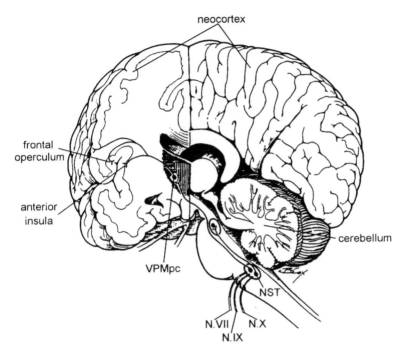

Figure 3.7   *Overview of the central taste system. Three cranial nerves (VII, IX, and X) bring taste input to the front part of the nucleus of the solitary tract (NST), while the back part of the nucleus receives visceral input. Taste signals proceed from NST to the thalamus (ventroposterior medial nucleus pars parvocellularis, or VPMpc), and then to the taste cortex in the frontal operculum and anterior insula. Drawing by Mr. Birck Cox. Adapted from Pritchard.[12]*

cortex, but there is also a loop from insula to amygdala to orbitofrontal cortex. The amygdala has itself proved to be an important relay for taste information, and perhaps for the motivational aspects of eating.[13] Orbitofrontal cortex also receives visual input from the visual cortex (where object recognition takes place), and olfactory information from the piriform cortex via the mediodorsal thalamus. All three modalities – vision, smell and taste – have been shown to impinge on individual cells within the orbitofrontal cortex, giving it the capacity to integrate a variety of sensations associated with the identification and assessment of foods.

Both the orbitofrontal cortex and amygdala project to the *hypothalamus*, which may be concerned with the motivational aspects of eating. In addition, several if not all of the forebrain taste areas – VPMpc, insular–opercular cortex, orbitofrontal cortex, amygdala and hypothalamus – project back to the NST, establishing a circuit through which higher-order functions, such as hunger levels, could influence the reflexive acceptance or rejection of foods.

NEURAL DEVELOPMENT OF THE CHEMICAL SENSES

Development of the olfactory and gustatory systems is a process that continues throughout life. Receptors, in both chemical senses, are formed early in foetal development, then undergo a constant cycle of degeneration and replacement. New receptors are created from the cells of the surrounding epithelium.

Olfaction

Development of olfactory receptors. Early in embryonic development, two oval patches can be seen at the front of the forming head. A ridge soon develops around these patches, creating a depression that is lined with epithelial tissue distinct from the surrounding epidermis. This is the *olfactory placode*, the birthplace of our sense of smell. The pit deepens at the expense of the underlying tissue that separates the placode from the pharynx, and finally breaks through to form the posterior naris, connecting the nasal cavity with the pharynx.

The acuity of sense of smell is related to the surface area over which the olfactory receptors are spread. To enlarge that surface, folds called *turbinates* grow within the nasal cavity. In smell-poor humans, there are three turbinates on the lateral wall of the cavity. Most mammals, however, have elaborate structures that offer an enormous surface area to accommodate a proportionately large number of receptors.

In addition, a secondary olfactory organ, the *vomeronasal organ*, forms as a pouch on the inside wall of the nasal cavity. Receptors in the vomeronasal organ detect pheromones – organic molecules that carry information meaningful to members of a species. In human embryos, the vomeronasal organ and its central target, the accessory olfactory bulb, both develop during the first trimester, but then regress and nearly disappear before birth.

Embryonic sensory cells develop early in gestation and, by the ninth week in humans, they form dendritic knobs from which cilia begin to grow. On these are the receptor sites where transduction of olfactory information ultimately occurs. At the same time, axons from these cells grow in the opposite direction, towards the central nervous system, to invade the developing olfactory bulb, there to form synapses with second-order neurons.

Sensory cells become functional towards the end of gestation. They respond to all odorants at first, but soon pare their sensitivity to a subset of stimuli for which they henceforth bear the responsibility of detection. Humans, like all mammals, are born with an operating olfactory system, a requirement in many species for locating the nipples that offer survival.

Development of the olfactory bulb. The olfactory bulb begins to take form as the invading axons of the sensory nerve reach its location. In a pattern typical of mammalian development, the large output neurons (mitral cells) are born first, even before the structure of the bulb that will house them takes shape. Next are the mid-sized tufted cells, spawned in the central cavity of the developing bulb. They must migrate outwards, through the mitral cells, to take their positions in the external plexiform layer, in a process that presages the inside-out development of the cerebral cortex. Finally, the small periglomerular and granular cells are formed, to work their way out to the granular layer of the bulb.

The glomeruli begin forming late in gestation, and continue to take shape post-natally. They are organized under the influence of the approaching sensory axons which, in fact, will create a glomerular structure in any neural tissue, olfactory bulb or not. However, when the tissue is fraudulent, imported from some other area of the brain, the resulting glomeruli will not be functional and the animal will have no sense of smell.

Taste

Development of gustatory receptors. The tongue develops from the primitive pharynx. It has two sections: the anterior oral region and the posterior pharyngeal area. The oral region of the human tongue begins to develop at four weeks gestation, and by seven weeks has assumed its adult form. Further development involves enlargement and differentiation of the musculature.

Papillae. Development of the taste receptors proceeds from large structures to small: from papillae to buds to taste cells (*cf.* section on Gustatory Anatomy). All three types of taste papillae take form in the human between 8 and 9 weeks of gestation, and achieve their adult structures well before birth. Fungiform papillae appear as small, poorly defined protuberances on the front of the tongue. As they grow, they assume a raised position as the epithelial cells around them grow downwards. They are subsequently surrounded by filiform papillae, instrumental in carrying touch information from the tongue.

Circumvallate papillae in their early stages resemble large fungiforms. The surrounding epithelium then grows down and around the core of the papilla. Small cavities appear in the epithelium, which eventually fuse to form the moat that surrounds the circumvallate papilla. The downgrowth also stimulates creation of the *von Ebner* lingual salivary *glands* whose secretions drain into the clefts of the papilla.[14]

The developmental sequence of foliate papillae is not yet known.

Taste buds. The earliest signs of taste buds appear as the papillae take form. Groups of cells having staining characteristics different from those around them collect in the newly defined gustatory epithelium. They are covered by epithelial cells and have no communication with the surface of the tongue. Gradually, these cells elongate to span the thickness of the epithelium in which they are embedded. Then, at about 85 days gestation, a channel is formed in the overlying tissue to create the taste pore through which the developing cells of the bud sample the environment.

As in other developing neural systems, there is a prenatal overproduction of taste buds that occurs as the large buds divide into those that are smaller and more mature. Buds reach their maximum number at about the time of birth, in conjunction with a peak in the number of innervating peripheral nerve fibres that presumably create them. Then, in parallel with a decline in nerve fibres, taste buds are lost through competitive interactions.

Taste receptor cells. Taste cells are formed and functional by the third trimester. Recordings from the peripheral nerves of sheep foetuses reveal an immature pattern of responsiveness to taste stimuli – broad, poorly differentiated sensitivity to various taste qualities – but responsiveness nonetheless. Human foetuses have been observed to increase their rate of sucking *in utero* when saccharin was injected into the amniotic fluid. Thus, sweet tastes are not only detected by the foetus, they are rewarding and can motivate behaviour.

Peripheral taste nerves. Nerve fibres (axons) invade the basement membrane of papillae, there to stimulate epithelial cells to differentiate into taste buds. The number of buds is directly related to the number of axons, and all buds degenerate if the afferent nerve is severed, even in maturity. As the axons regenerate to the tongue, taste buds are reborn. The chorda tympani, which serves the front two-thirds of the tongue, is the first nerve to arrive. It projects both to the papillae and around them, but the surrounding fibres soon retract and leave the nerve focused on the taste buds. Only then do trigeminal fibres arrive at the tongue, to ascend the cores of the papillae along with the chorda tympani axons, and to innervate all regions except those already claimed by their predecessors. Thus, the anatomical distinction between taste and touch is established.

While the full complement of taste fibres is achieved early in development, these axons do not have a fully mature structure or function. Myelination proceeds well after birth. Its completion is associated with shorter response latencies and greater specificity to different taste qualities.

Development of the central taste system. The chorda tympani nerve derives from the *geniculate ganglion*; the glossopharyngeal from the *petrosal ganglion*; the vagus from the *nodose ganglion*. The geniculate is the first of all ganglia to show mature cells in the rat, and is followed closely by the other two. The central projections from the ganglion cells constitute the tractus solitarius, a pathway for gustatory and visceral information in the medulla. Differentiation is poor at first. Fibres from the geniculate ganglion enter the brain in concert with those from trigeminal and vestibular ganglia, carrying touch, temperature, pain and balance information. This common sensory tract eventually differentiates as the gustatory fibres proceed caudally to join those entering through the glossopharyngeal and vagus nerves in the NST, where second-order gustatory neurons reside.

Much dendritic growth continues for another three weeks, and the responsiveness of the neurons to taste stimuli continues to increase. It is, indeed, a general feature of the developing gustatory system that functional activity is robust even before the receptors or their central destinations are mature. Thus, there is opportunity for environmental stimuli to have a substantial impact on the mature character of the system.

Regeneration. Taste buds cannot be created or sustained without peripheral nerve innervation. If the taste nerves are cut, or if axonal transportation along their length is interrupted, taste buds degenerate within two weeks. Thus, there is some unidentified trophic factor that courses peripherally to induce the formation of buds and to maintain them. If the area of innervation of two cut taste nerves is reversed, each is perfectly adequate to form buds in its new location. Even non-gustatory fibres that originate from taste ganglia will do so. Other sensory nerves, however, have no such capacity.

Moreover, the innervation must be at a site on the tongue that is prepared to receive it. A taste nerve joined to tongue epithelium at locations devoid of papillae will have no effect, nor can it induce bud formation at all on a tongue from which the papillae have been removed. Therefore, the taste fibres are unique among sensory nerves in possessing a chemical to which only cells contained within papillae on the tongue will respond. This is fitting for a system that is in a constant state of regeneration throughout its life.

RECEPTOR MECHANISMS

The generation of chemosensory information relies on the dissolution of molecules in either a gaseous medium into the viscous mucus of the olfactory epithelium, or in a liquid medium into the saliva. In both cases, the cilia (olfaction) or microvilli (taste) of receptor neurons extend into the receptor surface area to provide channels for transduction. Movement from one phase to another favours molecules with certain properties and excludes others. Therefore the effectiveness of transduction depends on a sequence of

Table 3.1 *Comparison of the elements associated with the transduction sequence in olfaction and taste. Adapted from Kinnamon and Getchell.*[15]

Sequence	Olfaction	Taste
Stimuli	Organic molecules Primarily hydrophobic Primarily volatile	Electrolytes: organic molecules Primarily hydrophilic (except bitter) Primarily non-volatile
Submodalities	Several to many	Sweet, salty, sour, bitter, umami, metallic, *etc.*
Access	Free diffusion (hydrophilic) Transport molecules (hydrophobic)	Free diffusion (hydrophilic) Possible transport (*e.g.* bitter)
Perireceptor environs	Mucus	Saliva and mucus
Receptor cell organization	Olfactory receptor neurons dispersed in epithelial sheet	Taste cells located in taste buds
Transducing element	Cilia	Microvilli
Stimulus detection	Receptor molecules	Receptor molecules Apically located ion channels Ligand-gated channels
Transduction	G proteins and second messengers Direct modulation of ion channels	Direct modulation of ion channels (salty, sour) G proteins and second messengers (sweet, bitter, amino acid)
Receptor current	Amiloride-sensitive cation conductance	Amiloride-sensitive Na^+ conductance (sweet, salty) Less K^+ conductance (sour, sweet) No conductance change (bitter)
Receptor potential	Depolarization Hyperpolarization	Depolarization and action potential Possible hyperpolarization
Spatial distribution, voltage-gated conductance	Na^+, K^+ channels: axon	K^+ channels: apical membrane Na^+, Ca^{2+} channels: entire cell membrane
Action potential	Axonal transmission	Local
Synapses	Chemical transmission Olfactory axons with mitral, tufted, and periglomerular cells	Chemical between taste cell and primary afferent; between taste cell, basal cell, and primary afferent
Transmitter	Putative: carnosine	Putative: acetylcholine, serotonin, noradrenaline (norepinephrine), peptides

events: stimulus access to the receptor sites, the level of recognition once that access has been gained, the electrical properties of receptor cells and the effectiveness of synaptic transmission to the peripheral nerves. These elements of signal transduction are summarized in Table 3.1.

Olfaction

Stimulus access. Odorants must pass through two physical phases to reach receptor cells. First, airborne molecules negotiate the convoluted chambers of the nasal cavity in the air phase to reach the olfactory mucosa. Then they must be dissolved and passed through the mucus to the receptors. Odorants partition at the air–mucus boundary according to their water solubility, defined thermodynamically by the air–water partition coefficient. Partitioning favours the water–mucus phase by a factor of ten for even hydrophobic molecules, such as musks. For strongly hydrophilic chemicals (*e.g.* pyrroline), the factor rises to 100 000. In each case, the concentration of odorants is amplified, but the degree of amplification is the first factor in determining our sensitivity to a molecule. For free diffusion, hydrophilia is favoured.

Once in the liquid phase, the access of a molecule to the olfactory cilium where transduction begins is regulated by the physical and chemical properties of mucus. In humans, the viscosity of mucus may vary by a factor of 20 000, from that of water to tracheal mucus, according to its glycoprotein and water content. Mucus also contains three types of proteins that influence olfactory transduction. These are the membrane-bound odorant receptors at whose sites transduction occurs, *odorant-binding proteins* (OBPs) that favour the transport of certain molecules over others and enzymes whose functions are not well defined. OBP composes about 1% of nasal protein, and binds lipophilic odorants with a wide variety of molecular structures, offering them access to receptors that they would otherwise be denied by their hydrophobicity. Thus, olfactory mucus has physical properties and chemical constituents that may serve as aids or barriers to olfactory sensitivity. Its composition reflects the pharmacological condition of the individual, and so may vary with sex, age, health status, or reproductive state.

Transduction. Receptor cells presented with odorants through the processes described above must encode the quality and concentration of these molecules, and transmit the information to the olfactory bulb with high fidelity.

Transduction occurs at sites on the cilia of receptor cells that are studded with transmembrane receptor proteins. Those parts of the protein that extend beyond the receptor cell membrane are invested with *glycoprotein receptive sites* for the recognition of the thousands of odorant molecules that compose our olfactory perceptual experience. On the intracellular side of

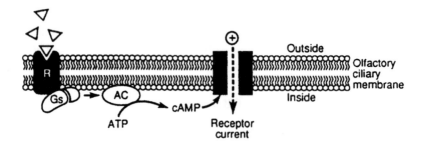

Figure 3.8   *Sequences in the transduction process when odorants are received by olfactory
receptors. The receptor (R) activates a guanosine triphosphate-binding stimulatory
protein (Gs), permitting it to couple with adenylate cyclase (AC). This coupling
initiates an enzymatic cascade that breaks adenosine triphosphate (ATP) down into
cyclic adenosine 3',5'-monophosphate (cAMP) which opens the ion channel,
permitting cations (+) to flow into the cell. The influx of cations depolarizes the cell
and causes the receptor potential that leads to an olfactory perception. From
Kinnamon and Getchell.[15]*

each receptor is a *guanosine triphosphate (GTP)-binding protein* whose role is to
amplify the effects of transduction; those specific to the olfactory system are
designated G_{olf}. A single coupling of odorant to receptor site activates
hundreds of G_{olf} proteins. When the activated G_{olf} binds with *adenylate cyclase*
(AC), found within the receptor cell, it forms the basis for reducing adenosine
triphosphate (ATP) to *cyclic adenosine 3',5'-monophosphate* (cAMP). It is cAMP
that opens the channels through which sodium ions enter to depolarize the
receptor cell and create the action potentials that result in olfactory
perception (Figure 3.8). This entire process is referred to as a second-
messenger system, where cAMP is the second messenger, and G_{olf} and AC
are the intermediaries that enable its action.[16]

Action potentials. Olfactory receptor cells are small and have a membrane
potential of about −55 mV. Receptor currents, created by sodium ion influx,
flow passively to the *axon hillock*, where they activate Na^+ and K^+ ion channels
to initiate an action potential. Axons of receptor cells are small (0.2 μm) and
unmyelinated, and are unusually impoverished of Na^+ channels (3/ μm²), all
of which conspires to limit their conduction velocity to about 0.17 m/sec,
among the slowest in the human body. Consequently, it requires more than
30 msec for the signal to traverse the 6 mm distance from the receptor
epithelium to its first central synapse in the olfactory bulb.

   Once there, receptor cell axons converge on the mitral and tufted cells and
on short-axon periglomerular cells of the glomeruli (*cf.* Olfactory Anatomy
above). The degree of convergence – and hence potential amplification – is
extraordinary: 15 000 primary axons impinge on a single mitral cell. The
main neurotransmitter that mediates this interaction is *glutamate*, whose

primary effects are probably modulated by *carnosine* (a dipeptide: β-alanyl-L-histidine) (Table 3.1).

Taste

Stimulus access. In contrast to odorants, most taste stimuli are non-volatile and water soluble. This includes acids (sour), salts (salty and bitter), sugars (sweet), amino acids (bitter, sweet and umami) and proteins (bitter and sweet). These molecules are released during chewing, and gain access to taste receptors by passing through the saliva and mucus that overlie them. Alkaloids (bitter) and a few other stimuli are hydrophobic. Despite the implied difficulty these molecules should have in traversing the aqueous medium, they are typically perceived at very low concentrations. This suggests the existence of a special transport system to ferry them to receptive sites on the microvilli.

Saliva is a broth of water, electrolytes, proteins and mucus secreted by the major salivary glands of the mouth – parotid, submandibular, sublingual – and by von Ebner's glands deep in the foliate and circumvallate buds. Secretions of the major glands bathe the entire oral cavity and create the global chemical environment in which taste transduction occurs. Von Ebner's glands pour their secretions directly into the clefts of foliate and circumvallate buds, and may provide the protein that transports hydrophobic molecules to their destinations. An 18 kDa protein with a striking sequence homology to that of OBP has been identified in von Ebner secretions, and dubbed VEG. It is supposed, although not yet demonstrated, that VEG works in a manner analogous to OBP in olfaction to transport critical molecules that would not otherwise have access to receptor sites.

Saliva plays a number of roles in taste perception. First, Na^+ and Ca^{2+} cations serve as carriers of the depolarizing potentials that initiate the taste signal. Secondly, the concentrations of electrolytes influence taste thresholds, setting background levels against which taste stimuli are detected. This pertains especially to Na^+, whose salivary concentration of about 2 mmol sets the normal absolute threshold for detection of NaCl, and the HCO_3^-, which is released as a buffer upon the detection of acids, helping to restore pH and maintain acid sensitivity. Finally, saliva contains enzymes that join with the physical action of chewing to begin the digestive process.

Mucus appears to be a contribution of the taste buds themselves. It may serve to cleanse the taste pore region, and to influence the access of molecules to receptor sites by direct chemical alteration or by causing differential rates of diffusion to the microvilli.

Transduction. Taste receptors exist in the form of ion channels or specialized receptor proteins on the microvilli of taste cells. The interaction of tastants at these sites leads to a change in membrane conductance, and to depolarization of the cell. The depolarization is conveyed passively from the top to the bottom of the taste cell, where it is responsible for transmitter release and initiation of an action potential in the nerve.

Both the receptor and the transduction processes vary according to the stimulus molecule. Salty, sour and most bitter stimuli interact directly with the ion channels on the microvilli. Sweet stimuli, amino acids and at least one bitter compound (denatonium) require specific protein receptors. The former group of stimuli causes depolarization directly through ion channels, whereas the latter uses GTP-binding proteins and second messengers. In each case, the final common path for transmitter release requires an increase in intracellular Ca^{2+}, either by an influx through Ca^{2+} channels or by the release of sequestered Ca^{2+} from within the cell itself. How the central nervous system differentiates among tastants each of which causes depolarization of a receptor cell remains a mystery, unless one assumes that activity in a particular cell is interpreted as a response to a particular taste quality. This approach is called the labelled-line theory of gustatory neural coding because each axon in the peripheral nerve is interpreted as conveying only one basic taste, which provides its label.

The investigation of receptor function has presumed the existence of five basic tastes. Transduction mechanisms for salty, sour, sweet, umami and bitter stimuli have been studied independently, despite the fact that individual receptor cells typically respond to a variety of taste stimuli. Each cell, then, is presumed to maintain sites for multiple transduction processes on its microvilli, with the proportion of sites determining its response profile. The proposed mechanism for each class of tastant is represented in Figure 3.9.

Salty. The transduction mechanism for Na^+, Li^+, and K^+ involves simply the passive movement of these ions along their concentration gradients from the mucosal surface into the cell.[17] The resulting depolarization may spread passively to the basolateral portion of the cell, to induce an increase in Ca^{2+} and so initiate synaptic transmission.

Sour. The perception of sour depends to a large degree on the concentration of hydrogen ions in the bathing medium. However, there are other considerations. First, organic acids are perceived to be more sour than inorganic acids of corresponding pH. Then, within each class of acids, the effects of the anion on the resulting taste perception can be considerable.

The transduction mechanism for sour taste is becoming clear. At rest, there is a small but significant outward current from the apical membrane, carried by K^+. As the H^+ concentration rises, it increasingly blocks the K^+ channel, reducing this current, causing a decrease in K^+ conductance, restricting K^+ to the cell, and so resulting in receptor depolarization as the cations build up.[18]

Acid-induced receptor potentials also decline in the absence of Ca^{2+}. This

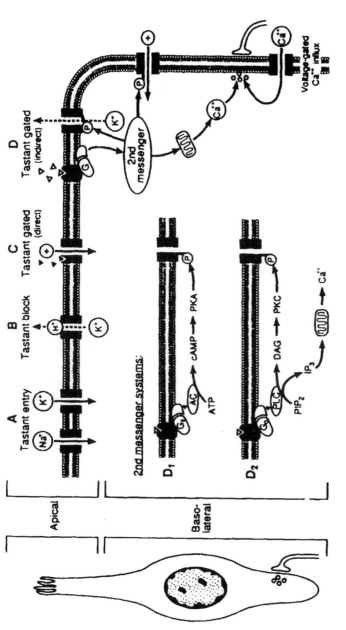

Figure 3.9 A schematic summary of the transduction mechanisms used in taste. (A): monovalent salts enter cells directly through ion channels, causing depolarization (salty). (B): Hydrogen ions block the efflux of potassium ions, confining them to the cell and causing depolarization (sour). (C): An amino acid gates an ion channel directly, such that its activation of the protein receptor allows cations to enter the cell, depolarizing it (umami). (D): Sugars and some alkaloids bind to protein receptors that are coupled to second messenger systems (sweet and bitter). The second messengers may be adenylate cyclase (AC) as in D_1, or Phospholipase C (PLC) as in D_2. From Kinnamon and Getchell.[15]

may implicate Ca^{2+} conductance as a second receptor mechanism, or it may merely reflect the fact that K^+ conductance depends partly on the presence of Ca^{2+}, without which K^+ is less activated, and so less subject to H^+ blockade.

Sweet. Only a minute proportion of stimuli are sweet, yet these comprise a remarkable diversity of molecular structures. A common physical property of these molecules is the possession of at least one site where the anions are separated by 3 Å, with one of them bound to a H^+. It is hypothesized that such a molecule interacts with a complementary complex on a sweet receptor site to form a pair of hydrogen bonds – the requisite physical condition for sweetness.[19] While this complex can be identified on all molecules that evoke a sweet sensation, its presence is no guarantee of sweetness. Thus, the theory lacks predictive power, so its utility is limited.

On the receptor side, protein fractions that undergo conformational changes in the presence of sweet-tasting molecules have been derived from the tongues of cows, rats and monkeys. However, they also bind molecules that are not sweet and have been found on non-gustatory epithelium, raising doubts about their status as the transduction sites for sweetness.[20]

While humans report sweet sensations to a variety of compounds, an individual's sensitivity to one class is not predictive of sensitivity to others. Moreover, if a person is adapted to one sweet stimulus, other chemicals can still elicit undiminished sensations of sweetness. These findings, along with the wide variations in sensitivity to sweet compounds across species, imply the existence of multiple receptor sites for sweetness. These are probably staffed by a series of specific receptor proteins, yet to be discovered.

Whatever the sites, the transduction of sweet tastes requires the presence of GTP, thus implicating GTP-binding proteins as part of a second messenger system. With binding, adenylate cyclase activity is stimulated, allowing cAMP-dependent phosphorylation, and so blockade, of K^+ channels. Such a blockade would lead to depolarization of receptor cells with an increase in membrane resistance – just the outcome that accompanies the application of saccharides to taste cells.

Bitter. The chemical requirements for bitterness are as general as those for sweetness are specific. There must be multiple transduction mechanisms to account for the variety of chemical structures that are effective in evoking bitter sensations. The hydrophobicity of many bitter molecules suggests a non-specific recognition mechanism. After being transported through the aqueous saliva and mucus by the VEG protein, the molecules may be absorbed within the lipid layer of the apical membrane, there to alter surface potential or activate a second messenger system. The degree of lipophilicity (hydrophobicity) has been shown to correlate with the intensity of bitterness for a wide variety of compounds, but does not satisfy all observations. The genetic trait of taste 'blindness' to thiourea compounds implies the existence of a specific molecule for the recognition of at least this class of bitter substances.

While the molecular structures and binding mechanisms associated with bitter taste transduction are unclear, the subsequent step – generation of a receptor potential – is better defined. First, bitter alkaloids, such as quinine, elicit depolarization accompanied by an increase in membrane resistance, the signal that K^+ is being blocked. Secondly, depolarization has been shown to result from an efflux of Cl^- from taste cells, an event that would make the intracellular environment more positive. Finally, denatonium chloride, the most bitter substance known to humans, works through an unidentified second messenger system to release intracellular Ca^{2+} that had been sequestered in the taste cell, and so to elicit a transmitter release.[21] The relative contributions of these three mechanisms may vary with the bitter stimulus.

Umami. Amino acids may induce a receptor potential by binding directly to protein receptors that are coupled to cation channels. Thus, a conformational change in the receptor protein would permit Na^+ entry, and so depolarization of the cell.

Since the processes by which tastants are recognized and coded are not fully defined, they may yet prove not to be as independent of one another as they now appear to be. It is doubtless, however, that the receptor mechanisms associated with various categories of taste qualities are rather distinct, certainly more so than those for olfaction.

Action potentials. Taste receptors communicate among themselves electrotonically, and so may operate as functional units of two or three cells. They form unusual synapses with afferent nerves, in that the polarity is not clearly specified. This implies that the afferent nerves may modulate the activity of receptor cells, even as they are receiving signals from them at the synapse.

In contrast to the single transmitter, carnosine, that appears to mediate the first-order communication in olfaction, taste lays claim to a host of potential transmitters: acetylcholine, serotonin, noradrenaline (norepinephrine), vasoactive intestinal peptide, substance P and calcitonin gene-related peptide have all been identified in the synapse between receptor cells and peripheral nerves of different species (*cf.* Table 3.1).

NEURAL CODING

Olfaction

Smell is much more than a sensory system. Its signals mediate arousal, memories, emotions, motivations and decisions to approach or avoid an object. It codes stimuli under the simultaneous demands of these diverse functions, and hence that code is complex and poorly understood.

Functional anatomy. The processing of olfactory information proceeds in four stages: first at the receptor level, then at the glomerulus and the external plexiform layers of the bulb and finally in the primary olfactory cortex.

The anatomy gives some indication of olfactory function. Each olfactory bulb in the rodent has 2500 glomeruli, and receives some 50 million axons from the epithelium, for a convergence ratio of 20 000:1. The 20 000 axons that enter each glomerulus terminate on the branching dendrites of just 23 mitral, 70 tufted and 450 periglomerular cells. Thus, each mitral cell receives input from nearly 1000 receptors; each tufted cell from 300. The arrangement is reminiscent of that in the human visual system where some 125 million receptors converge, through five layers, onto about one million ganglion cells whose axons form the optic nerve.

After passing through the straits of the bulb, the olfactory system opens again. The primary central target of the lateral olfactory tract is the piriform cortex, which contains in excess of one million pyramidal cells. Therefore, each of the 50 000 axons that compose the lateral olfactory tract contacts at least 20 cortical cells. In fact, the extensive branching of mitral and tufted cell axons makes this ratio much higher. The analogy with vision holds again, as the optic nerve projects to millions of lateral geniculate neurons, and they to perhaps 1000 million cells in the visual cortex. The implication is that an enormous amount of processing must go on in the olfactory bulb to prepare the afferent signal to serve its several functions at the cortical level.[22]

The range of centrifugal inputs returning to the bulb from diverse regions of the brain reinforces this implication. The nervous system controls and adjusts sensory signals to extract the most vital information, and it does so mainly through centrifugal fibres. That so many brain areas project back to the bulb means that there is widespread interest in influencing the decisions that are made there. The only targets in the bulb are intrinsic neurons, *i.e.* those that do not project back out of the bulb but rather mediate local inhibitory interactions.

Characterization of the olfactory system. We have inherited our study of olfactory neural coding from the strategies used to define vision and hearing. The issues have not changed substantially in decades. First, what are the stimulus characteristics that determine the odour quality? Secondly, are there primary odour qualities, each independent of the others, that may be combined in various proportions to create any smell? Thirdly, are olfactory neurons divisible into functional neuron types, one corresponding to each quality, such that neurons within each type are functionally identical in their sensitivities? Fourthly, is the code for quality carried in the pattern of activity of all olfactory neurons, or is it restricted to a dedicated channel composed of cells most responsive to that quality? Finally, is there a *topographic organization* in smell whereby each odour is signalled by cells that are physically grouped together?

First, the dimension along which olfactory stimuli are organized is one of hedonics. Odorants are spontaneously declared to be appealing or unpleasant, an evaluation that fits well with the mission of smell to mediate approach–avoidance behaviour. It is undoubtedly the case that some odours are genetically programmed to arouse strong hedonic reactions and behavioural responses (*e.g.* predator), while others do so only after some measure of experience, especially early in life (such as recognition of mother or mate).[23]

Secondly, there is no evidence that primary odours exist. Odours fall into broad groups, *e.g.* floral, but the distinctions among categories are poorly defined. Moreover, no small group of odour qualities has been found that can be mixed to create all olfactory perceptions, as can be done in vision using the three primary colours. Thus, the olfactory universe can best be characterized as a continuum of stimuli, with none having privileged status.

Thirdly, it has not been possible to classify olfactory neurons into a small number of discrete types, as has been done for the four types of photoreceptors in the primate retina. Indeed, the genetic instructions have been identified for the production of several hundred different types of olfactory receptors, and it is anticipated that the number may reach into the thousands. So there may be 10 000 different types of receptor cells, each represented 10 000 times over in the epithelium.

Fourthly, olfactory cells are broadly responsive. Any one receptor responds to a wide range of odour qualities and, conversely, an odour presented to the epithelium induces responsiveness in a high proportion of the cells. This implies that particular receptors are not dedicated to the recognition of specific stimuli, but rather that the code for each odour is carried in a pattern of activity across a large population of receptors.

Finally, there is some degree of topographic organization in the connections from receptor to bulb. Various odorants elicit their maximum responses in different parts of the olfactory epithelium. The distinctions are relative rather than absolute, for some level of activity is seen throughout the epithelium. Nonetheless, the characteristic distribution in response to a given odour offers the likelihood that a spatial code is one component of the signal for odour quality. This organization is preserved to some degree in the convergence of fibres from epithelium to bulb. Regions of the epithelium project densely to focal areas in the bulb, but also diffusely to surrounding areas. The process is such that the vague topographic organization of the epithelium is sharpened in the bulb, where any glomerulus processes information on only a subset of odour qualities. Moreover, this organization is not idiosyncratic to the individual, but is preserved across animals within a species and, indeed, across species. A particular odorant preferentially activates the same area of the bulb in mice, rats and rabbits.[24] The improvement in the precision of the spatial code requires that axons of receptors from different regions of the epithelium be guided to a focus in the bulb. The mechanism underlying this presumed process remains unknown.

Topographic organization is largely lost beyond the bulb. Vast regions of the bulb project to circumscribed foci in the olfactory cortex, while neurons from an individual glomerulus in the bulb are distributed across wide areas of the cortex. If the olfactory system can be viewed as addressing two questions – what is the odour, and how may its perception serve the multiple olfactory functions – then the bulb appears to have the best means of answering the first, and of distributing that answer to the most appropriate cortical area to address the second. Thus, the olfactory system appears capable of preferentially sending information with different types of biological relevance to those regions of the brain where it can be put to best use in controlling behaviour. In fish, for example, it has been shown that stimulation of each of a set of discrete regions in the lateral olfactory tract leading from the bulb evokes a specific and distinct behaviour. *On its input side, the bulb organizes chemicals; on its output side, behaviour.*

In sum, two opposite coding strategies are simultaneously operative in olfaction: responses are widely distributed, but that distribution is constrained by a degree of spatial organization. This implies that a specified region of the epithelium and bulb has primary responsibility for encoding a particular odour quality. Yet the distributed response suggests that the code within that region is carried in a pattern of activity across many neurons. Once the chemical analysis has been performed, the code assumes a different form, this time organized to elicit the proper behavioural response.

Transmitters in the bulb. The olfactory bulb is host to millions of afferent axons coming from the epithelium, to centrifugal inputs originating in a dozen regions of the brain, and carries out its functions using five distinctly different cell types. A variety of transmitters and neuroactive peptides, including nearly all those known to the central nervous system, are employed in the bulb to manage the resulting complex interactions.

Taste

The same set of five questions may be addressed to the taste system, with quite different answers.

Stimulus characteristics that determine taste quality. No issue is more fundamental to the definition of a sensory system than a determination of the stimulus characteristics upon which perceptions are based. The major dimension along which taste perceptions may be organized, however, relates not to any one feature of the stimulus molecule, but rather to a physiological characteristic: its effect on the health of the taster. Stimuli that provide nutrition are rigorously segregated from those that are poisonous.[25] Moreover, the analysis of whether a stimulus is nutritious or toxic is spread throughout the nervous system and serves to (1) control somatic reflexes through which immediate action may be taken to swallow or expel, (2) activate parasympathetic reflexes to anticipate and aid the digestive process, (3) arouse pleasure that sustains eating, or disgust that terminates it and (4) form a cognitive impression of the quality and intensity of the stimulus (*cf.* section on The Control of Eating).

The recognition that taste stimuli are organized according to their impact on our health should not be surprising. Natural selection among foragers, faced with a chemical world of a few nutrients among many toxins, favours the taste system that activates the appropriate reflexes to ingest or expel, and the hedonic tone of attraction or revulsion. Those ancient creatures who delighted in the taste of toxins left few offspring.

The evolution of a system designed to distinguish the beneficial from the harmful is not unique to taste. Vision and hearing are used for the detection of predator and prey, and so promote escape from the former and capture of the latter. But predator and prey are not the organizing principles of those systems, as nutrients and toxins are for taste. For this is the primary function of taste: not to provide a continuous record of the surroundings, but to sample the chemical environment discretely, and to predict in each case the consequences of ingestion. The sense of taste looks not just beyond the body, but within it. Its unifying organization is not physical, but physiological.

Taste primaries. Most of the history of taste research has been occupied with the development and generation of support for classification schemes. It was only occasionally questioned whether there *are* classes of tastes; the controversy usually surrounded their number and identity. While that number has ranged from two to eleven, classification systems have been dominated from the time of Aristotle to the present by four qualities: sweetness, saltiness, sourness and bitterness.

Evidence for the independence of four basic tastes comes from a variety of sources that range from patch clamp recordings to ecological considerations:

(1) Structure–activity relations. A distinct molecular structure and receptor mechanism is proposed for the transduction of each of the basic tastants, as described above.

(2) The neural code. The monkey's behavioural response to sugars parallels activity in a sweet channel as opposed to the total response evoked by each sugar. This implies that the monkey's perception of sweetness comes exclusively from one sensory channel.

(3) Topographic organization. The gustatory receptor surface is differentially sensitive to four basic taste stimuli – NaCl, sucrose, HCl and quinine. This spatial separation of qualities is maintained to some degree at central neural relays.

(4) Concentration–response functions. The rates at which the perceived intensities of the basic stimuli increase with concentration are distinct. HCl rises with the steepest slope, followed by sucrose, NaCl and quinine.

(5) Temporal properties. The time course of the response evoked by each basic taste is neurally and perceptually distinct.

(6) Temperature and taste. Gustatory thresholds change with temperature, and the interaction is unique for each of the four basic tastes.

(7) Cross-adaptation. Adapting the tongue to any one of the basic tastes does not have a significant neural or perceptual effect on the others, implying independent coding mechanisms.

(8) Taste modifiers. There are several chemicals whose applications affect subsequent taste perceptions. The most notable are gymnemic acid (abolishes sweetness), miraculin (makes acids taste sweet) and amiloride (suppresses saltiness). Both the neural and perceptual effects of taste modifiers are restricted to the specified taste qualities.

(9) Ethological considerations. Each primary quality serves a distinct physiological function; namely, ensuring energy reserves (sweet), maintaining electrolyte balance (salt), guarding pH (sour, bitter) and avoiding toxins (bitter).

Much of the foregoing, however, comes from verbal reports that have sometimes incorporated cultural and linguistic biases. When these are minimized by the use of semantic differential scales,[26] the independence of the basic taste qualities becomes less clear. Thus, taken together, this series of arguments is persuasive but not definitive. It has convinced most researchers that accepting the notion of primary tastes offers the advantages of conceptual organization, while not grossly misrepresenting the gustatory system.

If primary tastes are acceptable as a concept, we may turn to their number. A subset of the arguments advanced above in support of the traditional quartet of qualities has been invoked to promote another candidate: umami (savouriness) as represented by monosodium glutamate (MSG). MSG is transduced by a specific receptor protein,[27] it elicits a pattern of neural activity that is distinct from those of the other basic stimuli and its taste generalizes only poorly to theirs, it is unaffected by taste-modifying agents and it does not cross-adapt with any basic taste.[28] Thus there is support from the level of receptor processes, through sensory coding to perception that umami is a taste quality of independent standing.

Whatever the outcome of the debate over umami, it is clear that the number of taste primaries is not settled. Other amino acids, proteins, nucleotides, lipids, vitamins or minerals may vie for primacy in selected species or across phyla.

Gustatory neuron types. The foregoing discussion of primary taste qualities relates to the perceptual organization of taste. Another issue is whether there are gustatory neuron types that compose the taste system which imposes that organization. Are all neurons generated from a small number of templates, each replicated many times, as in the four classes of primate photoreceptors? Or is each taste cell unique in its properties, taking its place along a continuum of taste-sensitive neurons, each differing slightly from the next, as in auditory receptors?

The history of taste coding has seen this issue apparently resolved, first in favour of neuron types, then against, now in favour again. In the first recordings from single afferent fibres, three neuron types were described.[29] Subsequent sampling of more cells, however, identified many whose sensitivity profiles did not conform to the established categories. Within two decades, the accumulated profiles had become so diverse that the notion of neuron types was replaced by that of a continuum along which neuronal sensitivities are regularly spaced.[30]

But taste neurons *are* amenable to classification according to their location, structure, connections, sensitivity patterns and functions. Sensitivity to taste quality is not uniformly distributed throughout the oral cavity. Thresholds to sweet and salty stimuli are lowest at the tip of the tongue, to acids on the posterior tongue, and to quinine on the soft palate. This arrangement is preserved to some degree at each successive gustatory relay through the cortex. Close inspection of the structures and connections of cells in the NST of hamsters reveals that some cells are elliptical, with their long axes in a horizontal orientation, while others are star-shaped with their long axes oriented vertically. The former project to higher-order taste neurons, presumably to mediate quality–intensity evaluations and the reward value of a tastant. The latter project to a variety of nuclei below and behind the NST, perhaps to control the somatic and visceral reflexes associated with different taste qualities.

There is a tendency for any taste cell to show one of three recurring profiles of sensitivity – sweet, salty or sour-bitter – in response to an array of stimuli. The cells within any one group do not have identical profiles, but they are highly similar, and distinct from the profiles of neurons in either of the other groups. Moreover, the implication that there are identifiable neuron types is reinforced by using a specific sodium channel blocker, amiloride, on the tongue to determine if it affects the various types of cells differentially. When sodium transduction was blocked, the response of NST cells to NaCl was suppressed among salt cells, but the smaller response to NaCl in sour–bitter neurons – presumably not related to the mediation of saltiness – was unaffected. The categorization of neurons provided nearly perfect predictability as to which cells would be affected by interfering with a transduction mechanism two synapses removed. Thus, there is a clear separation of coding responsibility according to gustatory neuron type. Since this segregation of function is maintained from the receptor to the NST, it may be considered evidence not just for gustatory neuron types, but for discrete information channels within the taste system.

The relevant neural population. In the preceding section we showed that taste cells may be categorized, and that the resulting groups may operate through independent channels, each of which bears responsibility for recognizing a basic stimulus category. Psychophysical data from cross-adaptation and taste mixture experiments, cited earlier in support of primary tastes, strengthen this argument for specific coding channels, known as *labelled-lines*.

The persistent difficulty facing the labelled-line theory is that most mammalian taste neurons are sensitive to a wide array of stimuli. The more broadly responsive a neuron is across the basic taste qualities, the larger the proportion of its total activity that is beyond the channel of which that cell is presumed to be a part. This proportion must be ascribed to noise.

No such problem is encountered by the competing *pattern theory*, which asserts that *all* neurons contribute to the quality code through their relative rates of activity. This theory contends that a stimulus elicits a certain rate of discharge from each of a population of cells, determining a pattern of activity that is unique to the quality and intensity of that stimulus and that serves as its neural representation. Taste quality is coded in the shape of the response profile, read across neurons in a specified order, while the cumulative discharge rate – the area under the profile – represents intensity. Since every neuron contributes one point to the profile, each is theoretically of equal importance in the neural code. The more similar the patterns evoked by two stimuli, the closer their taste qualities (Figure 3.10).

The appeal of pattern theory derives from several sources. Firstly, it incorporates the broad responsiveness that characterizes most taste cells as the cornerstone of the theory. Secondly, it permits all neurons to participate in the taste signal, a useful feature in a system that has so few cells. Thirdly, it has predictive power. Stimulus pairs that evoke highly correlated patterns are difficult to distinguish from one another, require greater time for discrimination, and cross-adapt readily.

Thus, there is strong evidence both for labelled lines within the taste system, each line responsible for coding a basic taste quality, and for profiles whose shapes, read across all taste cells, carry the quality code. The primary reason that resolution between these competing theories has been so elusive

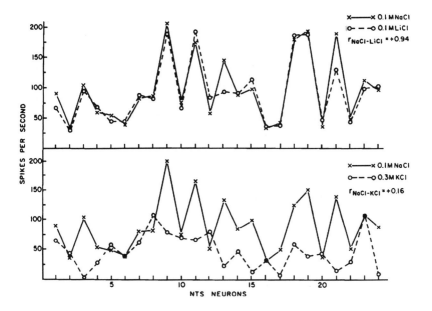

Figure 3.10 *The pattern of activity evoked from 24 cells in the nucleus of the solitary tract (NST) by NaCl, compared with the patterns elicited from the same cells by LiCl (top) and by KCl (bottom). The correlation coefficient between the patterns of NaCl and LiCl is +0.94, implying that these two salts have very similar tastes. The correlation between the patterns of NaCl and KCl is only +0.16, indicating quite distinct tastes for these two salts. From Doetsch and Erickson.[31]*

is that both can accommodate most data. Stimulus patterns become more highly correlated as a greater proportion of their evoked activity shares a common channel. While a patterning theorist may not recognize the existence of a channel, and a labelled-line proponent nothing *but* the channel, in fact the same set of neurons – channel or not – is generating most of the activity. Only with studies such as that with amiloride cited in the preceding section has it been possible to observe sharp distinctions across putative channels, and so to sway the balance towards the labelled-line hypothesis.

Alternatively, both theories may be operative, either in series or in parallel. In support of serial cooperation, there is evidence that the bold distinctions among basic taste qualities are made according to labelled-lines, while the subtle differences among stimuli within a category are mediated by the shape of the profiles they elicit within the labelled-line. The notion of parallel cooperation comes from studies of catfish. One set of taste fibres in this animal is exquisitely and specifically responsive to L-arginine. The second set is broadly sensitive to other amino acids, salts and acids. Thus, one group of cells may be devoted to recognizing a particularly salient molecule, while the remainder are deployed to report on the rest of the chemical environment.

Topographical organization. A simple and effective coding scheme, used in several sensory systems, is to separate neurons physically according to their functions. There are some indications that taste uses this strategy. There is differential sensitivity to tastants across the oral cavity, as detailed earlier, and this partial separation is maintained at each synaptic level of the taste processing chain. This has not been demonstrated clearly enough for it to be used as a major coding mechanism for taste quality. However, that evidence may now be at hand. Preliminary data from *functional magnetic resonance imaging (fMRI)* studies on humans suggest that taste is confined to the dominant hemisphere of the brain, and that there are discrete foci of activity in response to tasting NaCl versus sucrose.[32] If these data are confirmed, they will offer strong evidence that a person can evaluate a taste quality according to the location of the cortical cells that are activated.

THE CONTROL OF EATING

The value of the chemical senses is to guide the selection of foods. Feeding is laced with danger. Most organisms prefer not to be consumed, and defend themselves by physical or chemical means. While physical defences (*e.g.* the speed of the rabbit) are countered by physical weapons (the greater speed of the cheetah), chemical defences against poisons or hallucinogens must be detected by the chemical senses. Olfaction guides us to food and offers a strong suggestion as to its acceptability. Taste, poised at the threshold between the external and internal environments, makes the final decision to swallow or reject. The decision is complicated by the fact that eating, as opposed to drinking, places multiple demands on a person. Diverse nutrients are called for, and the mix among these may change with disease, age or nutritional or reproductive status. Yet humans and other omnivores can often select an adequate diet over time, an ability termed 'body wisdom' by Curt Richter.[33] To the degree that body wisdom is successful, the information needed to select a balanced diet comes partly from physiological signals from the gut, partly from past experience and partly from what the environment will provide. These diverse factors are reduced to food choices by the chemical senses.

Responses to Nutrients and Toxins

Taste information enters the central nervous system at the NST of the medulla (*cf.* section on Gustatory Anatomy). From here, at least four feeding-related functions are guided.

Somatic reflexes. Clusters of cells in the NST serve to orchestrate acceptance rejection reflexes. Each response – acceptance and swallowing or rejection and gagging – is fully integrated in the hindbrain, and is stereotypical for each of the basic tastes. Steiner[34] has studied the orofacial reflexes of normal adults, full-term and premature babies, anencephalic and hydroanencephalic neonates, blind and retarded adolescents, and patients with craniofacial abnormalities. To the degree permitted by their various limitations, all these people reacted similarly to the application of basic taste stimuli. The facial expressions served the dual purposes of dealing with the chemical (swallowing if nutritious, clearing the mouth if toxic), and of communicating an hedonic reaction to other people. The reflexes are innate and are functional by the seventh month of gestation.

Parasympathetic reflexes. A second projection from NST goes back to visceral areas of the medulla and to the dorsal motor nucleus of the vagus nerve, which controls many of the digestive processes. The digestive tract does not wait for a nutrient to be swallowed before responding to it. As soon as its taste is perceived, autonomic reflexes elicit insulin and gastric acid secretion and increase gastric motility to prepare the gut for food. These are called *cephalic phase reflexes* because they originate in the brain. A sweet taste evokes insulin release within seconds, but this preparatory response is blocked by cutting the vagus nerve that carries the command from medulla to pancreas. It is also blocked if the sweet stimulus had previously been paired with nausea, so creating a conditioned taste aversion to it. Under these conditions, the very signal for the formerly pleasant taste is altered to one more like that for quinine, and the cephalic insulin reflex is no longer engaged.

Discrimination of quality and intensity. A third projection from the NST goes forward to the thalamus, and then to the taste cortex. Electrophysiological recordings from alert macaque monkeys suggest that quality and intensity information is represented here. The evoked activity from the macaque cortex matched well with human reports of perceived intensity and with the similarity among a range of taste stimuli.[35]

Hedonic appreciation. From the primary taste cortex, projections go both to the secondary taste cortex (caudolateral orbitofrontal cortex) and to areas of the ventral forebrain associated with eating, most notably the amygdala and hypothalamus. Here, the pleasure of eating is presumably mediated. Whereas neurons in the primary taste cortex of a macaque are unaffected by its level of satiety, those in the secondary taste cortex, amygdala and hypothalamus respond to sugar only when it is desirable. As the animal becomes satiated from consuming a large quantity, the taste of the sugar loses its capacity to activate these cells whose responses are associated with pleasure.[36]

Alterations in the Taste Signal

It should be clear, then, that the chemical senses are organized to perform a general differentiation of toxins from nutrients, and that this ability is genetically endowed through ancestors who foraged successfully in a chemically inhospitable environment. Moreover, this analysis is accomplished at primitive levels of the nervous system, and directly influences hindbrain somatic and autonomic reflexes, cortical discriminative processes and powerful mechanisms for pleasure or revulsion. But we need even more. Our physiological needs are in constant flux. In just hours or days, the definition of which substances are acceptable as food may change, as the dangers of malnutrition weigh against those of toxicity. A starving person will accept a wider variety of tastes than one who is full, and those tastes will be more pleasant. After only minutes of eating, the balance may reverse again. If taste and smell are to provide the information on which the decision to swallow or reject is based, their signals should be modifiable to reflect these changing needs. There is considerable evidence that taste activity can, indeed, be modified to accommodate the individual experiences and momentary physiological needs of these animals.

Alterations based on experience. Our experiences have a profound and lasting influence on the tastes we prefer. The taste of mother's milk establishes preferences in rats that persist into adulthood. Preferences also develop for a taste that is associated with nutritional repletion. Thus, the distinctive flavours in a cuisine come to be associated with the satisfying effects of the meal, and those flavours are subsequently preferred above all others. In this manner, the world's cultures maintain their distinctions partly by their choice and preparation of the foods that bind their members.

Conversely, if a distinctive taste precedes nausea, even by as much as 24 hours, a powerful aversion is developed that will cause us to avoid that taste indefinitely. The reaction to the taste is not just dislike, it is revulsion. This is termed a *conditioned taste aversion (CTA)*, and most people have at least one. Note that the aversion is not created to other aspects of the person's experience: who dined across the table, the colour of the candles or the music in the background. There is a special relationship between the state of the gut and the one sensory system that makes the final decision about what will enter that gut – taste.

A rat may be offered sugar, followed by an injection that induces nausea. It will blame the illness on the sugar and develop a CTA. If the taste code representing that sugar in the nervous system is monitored, a change is found from a sweet signal to one that more closely approximates the signal for a bitter poison. Accordingly, the somatic reflexes to the taste of sugar reverse from acceptance to rejection, the digestive reflexes that would have led to a cephalic phase insulin release are blocked, the neurochemical signal for pleasure (*dopamine* release in the *limbic system*) is inhibited and the animal's overall behavioural reaction switches from approach to avoidance. The consequences of replacing a 'sweet' with a 'bitter' signal are manifested throughout all the components of eating.

Alteration based on physiological need. The momentary nutritional status of a person has a major impact on what foods are chosen. One constant and innate preference is for sodium, which we consume in excess when it is available. This preference becomes exaggerated when we are salt-deprived. Humans depleted by pathological states or by experimental manipulations show a pronounced craving for salt. Similarly in rats, whose neural responses to NaCl may be monitored when they are in sodium balance and following salt deprivation to determine how they change. It is found that taste neurons which normally respond to salt fall quiet during depletion, while neurons that usually respond to sugars begin signalling the presence of salt. Therefore, salt temporarily gains access to a channel that is normally reserved for only the most attractive tastes, making it attractive itself and supporting the powerful salt preference seen during salt deficits. Accordingly, the somatic reflexes that would normally serve to reject concentrated salt change to acceptance, the neurochemical signal for pleasure (dopamine release in the limbic system) is activated and the animal's demeanour towards strong salt reverses from avoidance to acceptance. Therefore, in both the negative (CTA) and positive (sodium appetite) cases, changes in taste activity in the hindbrain are sufficient to account for the resulting eating behaviour and for the reward derived from it.

While the appreciation of a sodium deficiency occurs over a period of days, the availability of glucose is of almost hourly concern. As blood glucose levels decline, we become hungry and foods taste better to us. As they rise, we feel satiated and foods lose their appeal. These effects are also manifested in changes in the taste signal. The activity of neurons in the NST of an alert macaque can be monitored as the monkey voluntarily feeds itself a strong sugar solution (Figure 3.11). Satiety is measured by the monkey's behaviour, its facial expressions and tongue movements, and its willingness to accept more sugar (bottom of each frame). As the animal progresses from acceptance to rejection, the responsiveness of NST neurons to the taste of the sugar does not change (top of each frame). The same approach was used in the primary taste cortex, with the same results. However, the situation changed when neurons in the monkey's secondary taste cortex were studied (Figure 3.12). Taste cells gave vigorous responses to sugar when the monkey was hungry. Then, as satiety increased, and acceptance turned to rejection, responsiveness declined to near spontaneous activity. That loss of sensitivity was maintained throughout the brain regions associated with the control of

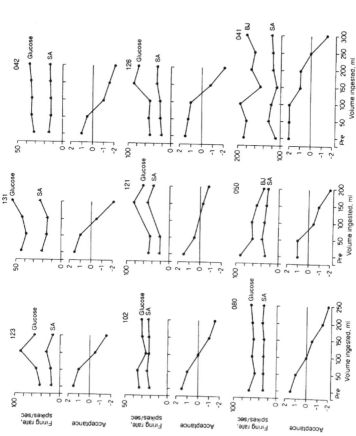

Figure 3.11 Spontaneous activity (SA) and neural responses evoked from cells in the nucleus of the solitary tract (NST) by the taste solution on which the monkey was fed to satiety. Each graph represents the results of a separate experiment during which the monkey consumed the sugar solution in 50-ml aliquots, as labelled on the abscissa. Represented below the neural response data for each experiment is the behavioural measure of acceptance of the solution on a scale of 2.0 (avid acceptance) to −2.0 (active rejection). The satiating solution is labelled on each graph. BJ – blackcurrant juice. Responses in the NST were not affected by increasing satiety. From *Yaxley et al.*[37]

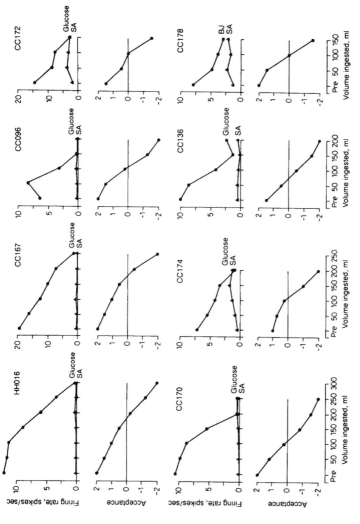

Figure 3.12 Same format as in Figure 3.11, except that responses are derived from single neurons in the caudolateral orbitofrontal cortex of the monkey (OFC). At this level of processing, discharge rate is a function of level of satiety rather than the purely sensory aspects of the stimulus. The letters and numbers in each frame identify the monkey and the recording track, respectively. From Rolls et al.[36]

appetite, most notably the amygdala and hypothalamus. Moreover, it applied only to the sugar that the monkey had been drinking. The responses to other appealing foods, such as peanuts, were only slightly affected. Thus, a person may be 'full' at the end of a main course, yet take renewed pleasure at the prospect of dessert.[35]

Thus, the sense of taste monitors both the external and internal environments to control feeding. Its external face must detect and orchestrate the rejection of toxins while recognizing and accepting nutrients. The internal face monitors the effects of the accepted chemicals on the body and remains current with its needs. To carry out these functions, the taste system has connections with somatic reflexes for swallowing or rejection, with autonomic reflexes to facilitate digestion, with cortical areas for a cognitive assessment of the chemical and with cortical and limbic regions to generate a sense of pleasure or revulsion.

While the mechanisms that underlie these processes are just now yielding to experimental investigation, their very existence should offer no surprise. We are the descendants of animals who properly identified, were attracted to and competed successfully for carbohydrates, fats, proteins and salt. The more pleasure our ancestors derived from eating these substances, the more likely they were to survive to become our progenitors. Since the search for nutrients is the most primitive of all motives, the appeal of these foods may be our most basic pleasure. Cloaked though it may be in cultural trappings, the pure biological reward of eating nutrients is irresistible to many.

The plasticity of the system permits finer adjustments. The acceptability of a food must be verified by its effects in the gut. If a taste that causes swallowing and pleasure is followed by nausea, the taste signal is changed – by as yet unknown mechanisms – towards one representing poison. Conversely, an inherently unappealing or neutral taste that is paired with nutrition gains appeal. Finally, within each individual, taste responses and their reflexive and hedonic effects are in flux to match the person's momentary physiological needs. The match may be unique and ongoing, as when a pathological state of sodium depletion leads to an intense and constant salt appetite.[38] More commonly, it reflects our normal eating patterns. When we are hungry, feeding reflexes are skewed towards the ingestive, and taste responsiveness to nutrients is heightened, as is the pleasure of eating. Fullness brings a suppression of the taste signal for reward for that taste, yet other chemicals may still arouse pleasure. It follows that a diet selected purely according to the pleasure derived from its components will be a varied one, perhaps skewed this way or that by the age or metabolic idiosyncracies of the individual. Therefore, a person has just one simple goal to fulfill: maximize pleasure. In most cases, what tastes best is what will best serve your physiological needs at the moment.

Why, then, are the chemical senses that offer this pleasure perceived in modern society as enemies to good health? Technical advances of the past few centuries have allowed industry to provide carbohydrates, fats, proteins and salt in unprecedented quantity and enormous variety. The orgy of consumption that followed has revealed a whole new set of pathologies that had never before played in our evolutionary development: cardiovascular stress, diabetes and hypertension. The chemical senses, tuned so exquisitely to the environment in which we evolved, clashed with the very society that had learned to pander to them. The intense biological pleasure of consumption has been dulled by the social disgrace of gluttony and by the deleterious effects on health that appear only at ages our ancestors never achieved. The physiologist's recognition that what tastes good is good for you has been reversed in the dieter's lament 'everything that tastes good is bad for you'.

REFERENCES

1. J. Garcia, D.J. Kimmeldorf and R.A. Koelling, *Science*, 1955, **122**, 157.
2. J.E. Amoore, J.W. Johnston and M. Rubin, *Sci. Am.*, 1964, **210**, 42.
3. B. Menco and A.L. Farbman, *J. Cell Sci.*, 1985, **78**, 283.
4. G.M. Shepherd and C.A. Greer. In *Synaptic Organization of the Brain*, ed. G. Shepherd, Oxford, New York, 1990, p. 133.
5. A.J. Pinching and T.P.S. Powell, *J. Cell Sci.*, 1971, **9**, 347.
6. L. Astic and D. Saucier, *Devl. Brain Res.*, 1982, **2**, 141.
7. I.J. Miller, Jr., *Anat. Rec.*, 1988, **216**, 474.
8. H.G. Schwartz and G. Weddell, *Brain*, 1938, **61**, 99.
9. R.B. Hamilton and R. Norgren, *J. Comp. Neurol.*, 1984, **222**, 560.
10. R.M. Beckstead and R. Norgren, *J. Comp. Neurol.*, 1979, **184**, 455.
11. R.M. Beckstead, J.H. Morse and R. Norgren, *J. Comp. Neurol.*, 1980, **190**, 259.
12. T.C. Pritchard. In *Taste and Smell in Health and Disease*, ed. T.V. Getchell, R.L. Doty, L.M. Bartoshuk and J.B. Snow, Raven, New York, 1991, p. 109.
13. T.R. Scott, Z. Karadi, Y. Oomura, H. Nishino, C.R. Plata-Salaman, L. Lenard and B.K. Giza, *J. Neurophysiol.*, 1993, **65**, 1810.
14. C.M. Mistretta. In *Taste and Smell in Health and Disease*, ed. T.V. Getchell, R.L. Doty, L.M. Bartoshuk and J.B. Snow, Raven, New York, 1991, 35.
15. S.C. Kinnamon and T.V. Getchell. In *Taste and Smell in Health and Disease*, ed. T.V. Getchell, R.L. Doty, L.M. Bartoshuk and J.B. Snow, Raven New York, 1991, 145.
16. R.R.H. Anholt, *Am. J. Physiol.*, 1989, **257**, C1043.
17. G.L. Heck, S. Mierson and J.A. DeSimone, *Science*, 1984, **233**, 403.
18. S.C. Kinnamon and S.D. Roper, *J. Gen. Physiol.*, 1988, **91**, 351.

19. R.S. Shallenberger and T.E. Acree, *Nature*, 1967, **216**, 480.
20. B.J. Striem, U. Pace, U. Zehavi, M. Naim and D. Lancet, *Biochem. J.*, 1989, **260**, 121.
21. M.H. Akabas, J. Dodd and Q. Al-Awqati, *Science*, 1988, **242**, 1047.
22. I.L. Kratskin. In *Handbook of Olfaction and Gustation*, ed. R.L. Doty, Dekker, New York, 1995, 103.
23. A. Holley. In *Taste and Smell in Health and Disease*, ed. T.V. Getchell, R.L. Doty, L.M. Bartoshuk and J.B. Snow, Raven, New York, 1991, 329.
24. F. Jourdan, *Brain Res.*, 1987, **417**, 1.
25. T.R. Scott and G.P. Mark, *Brain Res.*, 1987, **414**, 197.
26. S.S. Schiffman and R.P. Erickson, *Physiol. Behav.*, 1971, **7**, 617.
27. K. Kurihara, M.N. Kashiwayanagi, T. Nomura, K. Yoshii and T. Kumazawa. In *Chemical Senses*, ed. J.G. Brand, J.H. Teeter, R.H. Cagan and M.R. Kare, Dekker, New York, 1989, p. 55.
28. T.R. Scott and B.K. Giza, *Science*, 1990, **249**, 1585.
29. C. Pfaffmann, *J. Cell. Compar. Physiol.*, 1941, **17**, 243.
30. R.P. Erickson. In *Olfaction and Taste*, ed. Y. Zotterman, Pergamon, Oxford, 1963, p. 205.
31. G.S. Doetsch and R.P. Erickson, *J. Neurophysiol.*, 1970, **33**, 490.
32. J. Hirsch, R. de la Paz, N. Relkin, J. Victor, L. Bartoshuk, R. Norgren and T.C. Pritchard, *Chemorecep. Sci. Abstr.*, 1994, **16**, 314.
33. C.P. Richter, *Am. J. Physiol.*, 1936, **115**, 155.
34. J.E. Steiner. In *Advances in Child Development*, ed. H.W. Reese and L. Lipsett, Academic, New York, 1979, p. 257.
35. T.R. Scott, C.R. Plata-Salaman, V.L. Smith and B.K. Giza, *J. Neurophysiol.*, 1991, **65**, 76.
36. E.T. Rolls, Z.J. Sienkiewicz and S. Yaxley, *Eur. J. Neurosci.*, 1989, **1**, 53.
37. S. Yaxley, E.T. Rolls, Z.J. Sienkiewicz and T.R. Scott, *Brain Res.*, 1985, **347**, 85.
38. L. Wilkins and C.P. Richter, *J. Am. Med. Assoc.*, 1940, **114**, 866.

Chapter 4

Flavour Analysis

SUBJECTIVE VERSUS OBJECTIVE

The question is 'how should flavours be analysed?' There is no debate as to whether chemical analysis or sensory evaluation should be utilized to determine flavour products and the flavour quality of foods; both must be used. Sensory evaluation looks at the whole sample, is very reproducible and the analysis is usually done by averaging individual responses of trained judges. Objective chemical analysis is dependent upon sampling techniques and sample handling and/or separation techniques before measurement. But, as noted previously, some chemicals that cause large gas chromatography (GC) peaks have little or no odour. Thus, instrumental methods can only be utilized to measure flavours when they are calibrated by analytical sensory methods.

PSYCHOPHYSICS AND SENSORY EVALUATION

Psychophysics is the study of the relationship between the psychological perception of a stimulus and the physical stimulus that causes that perception. Sensory evaluation is the utilization of psychophysical techniques in the food industry to answer three types of questions:

(1) *Description*: What does the product taste like? How is one product different from another in quality? How do changes in the process, formulation, packaging, or storage conditions affect its perceived sensory characteristics?

(2) *Discrimination*: Would people notice the difference? How many people would detect it? If there is a difference, how great is it?

(3) *Affective or hedonics*: How much do people like this product? Is it an improvement over another product? Which attributes are liked or disliked? Which product would the consumer select?

The first consideration in any sensory study is to define the objective of the test. Do we need to know whether two samples are different in character (an analytical question), or which of the two is preferred (a subjective, consumer-oriented question)? The objective will determine the selection of panellists, the methodology employed, the appropriate level of statistical error to permit and the interpretation of results to provide a recommendation.

Panellists can vary from untrained consumers to specially trained flavourists. Target consumers must determine the preference for and acceptance of a finished product, whereas trained panellists are employed to differentiate among samples or to profile samples. All panellists, trained or not, must be screened for specific odour or taste deficits.

The instructions given to the panellists play an important role in any sensory test. The perceived sensory information, upon which memories of the event are based, is believed to be available for only a few tenths of a second. If not dealt with in that time, it is lost. Anticipated information is perceived more readily than that which is unexpected, and the panellist's motivation may influence what perceptions are retained. Information that passes through the selective filter of attention can be stored for seconds in short-term (working) memory, but if not rehearsed, it is lost. The limited duration of short-term memory dictates the number of questions that can be answered after one taste or smell of a sample.

Experiences during the panellist's life are an uncontrolled factor in testing. Familiarity with a stimulus will increase the likelihood of its recognition, and may bias the panellist towards a greater liking for it.

The sensitivity of the taste and olfactory systems varies enormously across individuals – sometimes by a factor of 100. Moreover, any individual's sensitivity also varies over time. Changes arise from ageing, illness, level of arousal, motivation, and expectancy. Body temperature alters taste and smell sensitivity because of blood flow changes across the tongue and olfactory epithelium, and this rises and falls with one's circadian rhythm. Sensitivity rises with the oestrogen surge of the menstrual cycle, and it rises when atmospheric pressure is low. In short, the psychophysical tests described below are subject to a host of variables, only some of which can be controlled by the sensory analyst, and so results must be interpreted statistically and with caution.

Discrimination Tests

Discrimination or difference testing is used to determine whether there is a perceptible difference between two or more products and, in some cases, the magnitude of the difference. As these tests involve comparative judgements, they can be very sensitive in determining small differences between products.

There are many types of difference tests: the paired comparison test, the triangle test and the duo–trio test are among the most commonly used in the food industry to measure the flavour differences in foods and food products.

Paired comparison. This test is used to determine whether two samples differ in a specific character; it is a directional test with a named attribute. For example, the panellist is presented with two samples and asked which one is more bitter. Positional bias is an important factor; thus, the sample order of presentation must be balanced so that there are equal numbers of AB and BA presentations. The statistical chance of obtaining the correct answer by guessing is 1 in 2, or 50%. Therefore, the detection threshold is normally placed at 75% correct responses, indicating that the difference has been detected on half the trials.

Multiple comparison. Efficiently used to evaluate four or five samples at a time, this test only measures the direction and magnitude of differences in one or perhaps two characteristics. A known standard is labelled as a reference or control and presented to the panellist along with several blind coded samples. The panellists are then asked to compare each coded sample to the known standard on the basis of an identified characteristic.

Triangle test. This test is used to determine an unspecified sensory difference between two products. The panellist is provided with three samples and told that two are the same and one is different. The objective of the test is to identify the different sample. In addition, the panellist is often asked to give the reason for his or her decision.

Taste order is specified because it has been shown to affect the test results: there is a tendency to select the middle sample as being odd. The six presentation possibilities (BAA, ABA, AAB, ABB, BAB, BBA), should be presented equally (the number of panellists being a multiple of six), or at least randomly. Because this is a *forced choice* test, that is a decision must be made even if no difference is perceived, there is a one-in-three possibility that the odd sample will be identified by chance alone.

The triangle test is popular because: (1) it is rapidly administered, (2) it is an easily understood task and (3) data analysis is simple.

Duo–trio test. This test is used to measure unspecified differences between samples. It is another three-sample difference test, where the panellist is first given a control sample and then asked which, of the next two samples, is the same as the control. The probability of guessing the correct answer in the duo–trio test is one in two, or 50%. This test is useful for panellist screening and training. Using immediate feedback, this practice helps teach panellists to discriminate, especially if allowed to retaste the samples and attempt to associate the similarities in the identical combination and then decide what is different in the odd sample.

Ranking test. This test is used to establish a magnitude of difference between samples on a specified attribute. The panellists are presented with 3–5 coded samples and asked to rank them in order according to a single specific attribute, *e.g.* sweetness. The taste order should be prescribed and a balanced design used.

Magnitude estimation. In this test, two or more coded samples are presented in a specified order, which is balanced across panellists. An arbitrary value for the attribute in question is assigned to the first sample. When panellists analyse (via tasting or smelling) the next sample, they assign a higher or lower value to it according to their estimate of the magnitude of the difference. Scales used could be *category scales, line scales* or *ratio scales.*

The simplest means of evaluating perceived intensity is to have panellists assign each stimulus to one of several *categories*, ranging from 'no flavour' to 'very strong flavour'.

The categories can be replaced with a *line scale*, along which the panellist places a mark corresponding to the intensity of the odorant. This takes more training, as standard concentrations of the attribute to be determined are used as anchors to define the ends of the line.

When a highly trained panellist assigns a series of numbers that represent the *ratio* of the perceived intensities of the presented flavourants to the first presented sample, a taste or odour's psychological magnitude can be related directly to its physical concentration by a logarithmic function. The logarithm of the magnitude estimation (P) is plotted as a function of the logarithm of stimulus concentration (C) as in equation (4.1),

$$\log P = n \log C + \log k \tag{4.1}$$

where k is the y-axis intercept and n is the slope. Taking the antilogarithm of each side yields a power function, equation (4.2),

$$P = kC^n \tag{4.2}$$

where n is the slope that relates psychological perception to physical concentration. The value of n is nearly always less than one for olfactory stimuli; hence a linear plot of perception versus concentration shows that our olfactory receptors become saturated (Figure 4.1).[1]

Magnitude estimation using the ratio scale offers a level of rigour not found in the other approaches, so its use is desirable as a psychophysical method. However, because of the following complications, it is rarely used in sensory testing in the food industry:

(1) To generate reliable numbers, the panellist must retain an accurate memory of the entire stimulus series.

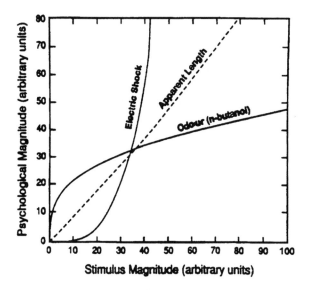

Figure 4.1 *Relationship between perceived magnitude of three types of stimuli, as measured by magnitude estimation and stimulus magnitude. Adapted and modified from Stevens.*[1] Copyright 1961 by MIT Press.

(2) If odours are presented too slowly, memory fades; if too quickly, adaptation reduces perceived intensity.

(3) The concept of ratios is unfamiliar to many people, and they use numbers as if they were categories, 10, 20, 30, *etc.*

(4) Finally, a number of non-sensory factors can affect the slope of the power function, including the range of numbers over which the ratings are made, the complexity and intensity range of the stimulant, and the order of stimulus presentation.

Threshold tests. A subset of difference testing is the measurement of the lowest concentration of a stimulus that a person can detect, or recognize. The *detection threshold* is the lowest concentration at which a stimulus is capable of producing a sensation, while the concentration at which the stimulus can be identified is called the *recognition threshold*. As described above, the recognition threshold is higher than the detection threshold. The *difference threshold* is the extent of change in the stimulus necessary to produce a noticeable difference. This has been labelled the JND for *just noticeable difference*.

These are not straightforward tests in olfaction, for the concentration of a molecule that actually reaches the olfactory receptors is a function not only of stimulus concentration, but also of solubility, volume, pressure and the respiratory cycle of the subject. In taste, saliva contains some sodium chloride, so the detection thresholds for salt will be elevated to just above the salivary concentration to which the panellist is adapted. Also, taste sensations are commonly confused at threshold levels.

The *method of constant stimuli* is the original measuring technique for thresholds. The panellist is presented with a series of randomized concentrations of flavour compounds ranging from imperceptible to clearly recognizable, and the point at which the correct response is given on 75% of the trials (*i.e.* halfway between the 50% chance level and 100% correct) is defined as the threshold. This approach suffers from the need for long times (at least several minutes) between sample presentations to ensure that adaptation effects have dissipated. With the requirement that there be several trials of each concentration, subject fatigue and boredom begin to influence the results. However, the method of constant stimuli retains the advantage of providing not just the threshold value, but also information about detection across a range of concentrations, so it is still used when time is not a factor.

The *method of limits* is in more common use today. The classic approach was to present alternating ascending and descending concentration series, and to take the mean of the transition points as the detection threshold. However, concerns about adaptation during the descending series have led to its deletion from most procedures, leaving only an ascending method of limits. This is efficient, but leads to a slightly higher estimate of threshold – *i.e.* poorer sensitivity – than does either the full method of limits or the method of constant stimuli.

The *staircase method* is a variant of the method of limits whereby an ascending series with large concentration jumps between steps is first given. This grossly determined threshold defines a tight concentration range to be explored intensively by either another ascending series (method of limits) or by a random presentation of concentrations (method of constant stimuli). Many passes may be made across this transition point while avoiding much of the adaptation that accompanies the higher concentrations. This two-stage method of first defining the approximate threshold, then focusing on a tight range around that point to provide greater definition, preserves the best features of both methods.

The approach for calculating the *difference threshold* (JND) is the same as the method of limits or the method of constant stimuli for determining detection or recognition thresholds, except that the comparison stimulus is an odorant rather than a blank. In sensory systems, the size of the change necessary for a JND is a constant proportion of the intensity from which that change is made. Thus, for example, a 10 decibel (db) tone may become perceptually louder when it is raised to 10.3 db, but a tone of 100 db must be raised to 103 db for a change to be detected. This is a statement of *Weber's Law*, which can be represented by equation (4.3),

$$\Delta I / I = K \qquad (4.3)$$

where I is the original physical intensity, ΔI is the change necessary to elicit a JND, and K is a constant. In the example of auditory intensity above, $K = 0.03$. The smaller the value of K, the greater the discriminative capacity of the sensory system. Values of K for olfaction fall between 0.15 and 0.35. The ability of humans to detect changes in stimulus intensity is not as advanced in olfaction as in vision or audition, yet it is not as impoverished as is often supposed. Only a few specialists who evaluate cosmetics, perfumes or wines ever have this capacity challenged, and so most people are unaware of their limits.[2]

Flavour units. In a particular food, it is useful to determine the concentration of an odour or taste, even if the chemicals responsible for those perceptions are unknown. The inverse of the determined dilution that produced the threshold level can be used as a flavour unit; for example, a sample having a threshold flavour of 2 ppm would contain 500 000 flavour units. The Scoville Heat Unit is a flavour unit used in the food industry to provide a quantitative measure of how much chilli pepper, black pepper or ginger was added to foods.

Descriptive Analysis

All descriptive analysis methods involve the identification, description and quantification of sensory aspects of a given product. They require trained personnel who are thoroughly familiar with the sensory attributes of the product being tested and who can accurately and precisely communicate their perceptions. The perceived sensory parameters which define the product are referred to by various terms such as attributes, characteristics, character notes, descriptive terms or descriptors. The choice of terminology for the sensory parameters is arbitrary, but must be agreed upon by all panellists during training sessions and applied in the same way during the testing. However, if the selected sensory attributes and corresponding definitions of these attributes can be related to the real chemical and physical properties of a product, the descriptive data are easier to interpret and more useful for decision making.

The intensity or quantitative aspect of a descriptive analysis expresses the degree to which each of the characteristics is present by assigning some value along a measurement scale (category, line or ratio). The validity and reliability of intensity measurements are dependent upon the selection of a scaling technique that is broad enough to encompass the full range of parameter intensities, but which has enough discrete points to pick up all the small differences in intensity between samples. Thorough training of the panellists is necessary so that they use the scale in a similar way across all samples and across time.

Descriptive analysis should be used to

(1) Define the sensory properties of a target product for new product development.
(2) Define the characteristics and specifications for a control or standard for quality assurance, quality control or research and development applications.
(3) Document a product's attributes before a consumer test to help in the selection of attributes to be included in the questionnaire and to help in explanation of the results of the questionnaire.
(4) Track a product's sensory changes over time with respect to understanding shelf-life problems.
(5) Map a product's perceived attributes for the purpose of relating them to instrumental, chemical or physical properties.

There are three commonly used forms of descriptive analysis and listings of descriptive terms: flavour profile method,[3] quantitative descriptive analysis,[4] and the spectrum method.[5]

Flavour profile method. This method, developed by Arthur D. Little, Inc in the late 1940s, involves the analysis of a product's perceived flavour characteristics, their intensities, order of appearance and after-taste. Four to six trained judges are together in the same room, but individually evaluate one sample at a time and record their results. Additional samples can be evaluated in the same session, but samples are not tasted back and forth. The results are reported to the panel leader, who then leads a general discussion of the panel to arrive at a 'consensus' profile for each sample. The data are generally reported in tabular form, although a graphic representation is possible.

Quantitative descriptive analysis (QDA) method. In response to dissatisfaction among sensory analysts with the lack of statistical treatment of data obtained with the flavour profile method, QDA was developed by the Tragon Corporation in collaboration with the Department of Food Science at the University of California at Davis in the 1970s. QDA panellists evaluate products one at a time in separate booths to reduce distraction and panellist interaction. The score-sheets are collected individually from the panellists and the data are entered into a computer. Panellists do not discuss data, terminology or samples after each taste session and must depend on the discretion of the panel leader for any information on their performance relative to other members of the panel and to any known differences between samples. The results of a QDA test are analysed statistically and the report generally contains graphic representation of the data in the form of a 'spider web' with a spoke for each attribute (Figure 4.2).

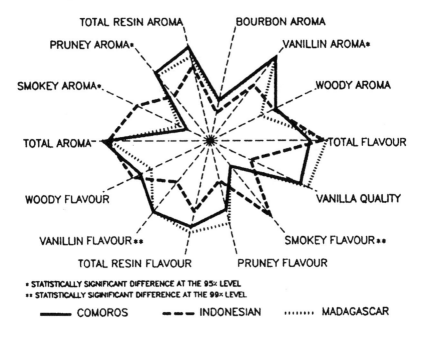

Figure 4.2 *A circular graph illustrating sensory differences in vanilla from different regions. From Gillette.*[6]

Spectrum method. This is a custom design approach to panel development, selection, training and maintenance developed by Sensory Spectrum, Inc. The aim is to choose the most practical system given the product in question, the overall sensory programme, the specific project objective(s) in developing a panel and the desired level of statistical treatment of the data. The method requires that terminology used as descriptors be developed and derived by a panel which has been exposed to the underlying technical principles of each modality to be described. The panellists must demonstrate a valid response to changes in ingredients and processing by choosing terms (and concrete reference samples for these terms) for flavours that vary with these changes. For quantitative differences in a characteristic, the method requires at least two and preferably 3–5 reference points distributed across the range to anchor the scale (category, line or ratio) used. A set of well-chosen reference points greatly reduces panel variability, allowing for comparison of data across time and products. The basic philosophy of the Spectrum method is to train the panel to define each product's sensory attributes fully, to rate the intensity of each and to include other relevant characteristics, such as changes over time, difference in the order of appearance of attributes and integrated total aroma and/or flavour impact.

Time-intensity techniques. As alluded to above, the perception of flavour occurs over real time: aroma, taste, texture and even thermal and pain sensations display dynamic changes in intensity over time. Most sensory analysis techniques have made the panellists estimate their responses before or after expectorating. But with the development of computers able to collect and process large amounts of data, important information such as the rate of onset of sensation, time and duration of maximum intensity and rate of decay of perceived intensity, time of extinction and total duration of the entire process can be made available. This increase in information has been useful in showing the time–intensity (T–I) differences in sweeteners judged to be identical in total intensity.[7]

To the panellist, the test is rather enjoyable. Using a computer with a lever (joystick or mouse) that can go up and down on an intensity scale, the panellist simply pushes up or down as the sensation being measured increases or decreases, while the computer monitors the time axis. A push-button can be provided to indicate the exact occurrence of such things as swallowing and chewing.[8]

Affective Tests

Also called consumer testing, affective or hedonic tests are used to assess the personal response (preference and/or acceptance) by current or potential customers. Most people today have participated in some form of consumer tests. Typically, a test involves 100 to 500 target consumers in three or four cities. Potential panellists are screened by telephone or in a shopping precinct. Those selected and willing are given samples, together with a score-card requesting their preferences and reasons for them, along with past buying habits, age, income, employment and ethnic background. Results are calculated in the form of preference scores overall and for various subgroups.

The most effective tests are based on protocols run among selected subjects of a target population that is anticipated to contain consumers of the specific products. The subjects or panellists are carefully selected to be representative of a larger population about which the investigator hopes to draw some conclusion. In the case of discrimination and descriptive tests, the investigator samples individuals with average or above-average abilities to detect differences. It is assumed that if these individuals cannot find a difference, the larger human population will be unable to detect it. In the case of affective tests, however, it is not sufficient merely to select from the general population. Instead, consumers of the type of product being tested must be recruited for the tests.

Consumer tests come in two forms. If a choice of one item over another is forced, then it is called a *preference test*. This test does not actually indicate whether any of the products are liked or disliked, only which is preferred; the researcher must have prior knowledge of the current or competitive product which is being tested against to make this judgement. In order to determine the 'affective status' of a product, one must perform an *acceptance test*, which is similar to attribute quantification in descriptive analysis except that the descriptor here is acceptance or liking. However, care must be excercised in the choice of questions asked and conclusions drawn; *e.g.* if one asks consumers if they like or dislike the sour taste, one does not know if they are also responding to their perception of bitter or astringent. Samples are usually rated along a hedonic category scale. Neither the numerical (linear) nor multiplicative (ratio) scale should be used, as they are too complicated for untrained consumers.

INSTRUMENTAL ANALYSIS

The reader is referred to the book *Introduction to the Chemical Analysis of Foods*, edited by S. S. Nielsen,[9] for a more complete discussion of food analysis and to the revised *Source Book of Flavours* edited by G. Reineccius,[10] for a more complete discussion of flavour-specific analysis of foods.

Wet Chemistry

As instruments have been refined, the analytical field has been retreating from the isolation and purification of flavour compounds. However, wet chemistry is sometimes necessary for the unequivocal identification of the chemical structure of a flavour. Large sample sizes are needed because of the low concentration of a flavour compound in a food. Primary separations are usually done via the traditional partitioning scheme (Figure 4.3) to separate acidic, basic and neutral flavour compounds from each other. This is accomplished by varying the hydrogen ion concentration of the water. Acidic compounds are extracted into the aqueous phase at pH 9–10; basic compounds are extracted into the aqueous phase at pH 3–4. Then secondary separations are performed chromatographically, such as with GC, liquid chromatography (LC) or thin-layer chromatography (TLC) on a preparative scale. A tertiary separation would use distillation for liquids and crystallization for solids. When isolated, the pure flavour compound is identified using at least two of the following techniques: mass spectrometry (MS), nuclear magnetic resonance (NMR), infrared (IR), ultraviolet–visible (UV/VIS) spectroscopies, and elemental analysis.

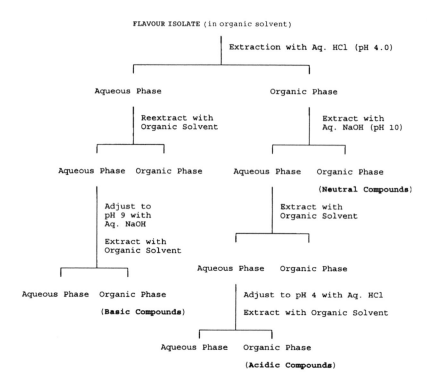

Figure 4.3 *Traditional liquid–liquid partitioning scheme for the separation of acidic, basic and neutral compounds.*

Taking advantage of the chemical reactivity and differences in functional groups of the individual classes of flavour components, separations may also involve liquid/liquid extraction (*e.g.* acids and bases), derivative formation (*e.g.* 2,4-dinitrophenylhydrazones of carbonyls) or complex formation (*e.g.* heavy metal binding to sulfur compounds).

Once the compound has been isolated and identified, quantitative analysis is used to measure intrinsic properties that vary with concentration and also have a minimum of interferences from other compounds in the sample matrix.

Absorption Spectroscopy (UV/VIS)

Absorption spectroscopy is based on the absorption of light at a specific wavelength in the ultraviolet (200–350 nm) or visible (350–700 nm) region of the electromagnetic spectrum. In order to determine the concentration of analyte in a given sample solution, the amount of light absorbed from a reference beam of light passing through the sample solution is measured. If the compound naturally absorbs the radiation, it is measured directly. When the analyte does not absorb radiation of the appropriate wavelength, it can be chemically modified during the analysis to convert it into a species that absorbs the appropriate radiation; this is called derivatization. In either case, the presence of analyte in the solution will affect the amount of radiation transmitted through the solution. Therefore, the relative transmittance, with and without the sample, may be used as an index of analyte concentration.

Absorption spectroscopy is a quick and easy means of analysing many flavours, as long as the sample is a liquid, translucent and not very complex. However, many foods are solid, opaque and chemically complex. When absorption spectroscopy is not appropriate, then a form of chromatography is usually used.

Chromatography

Chromatography is a general term applied to a wide variety of separation techniques based on the partitioning or distribution of a sample (solute) between a moving (mobile) phase and a fixed or stationary phase. It can be viewed as a series of equilibrations between the mobile and stationary phases. In GC, the mobile phase is a gas that flows through the stationary phase, which could be a solid or a liquid coated onto a solid matrix. The separation of compounds is determined by the relative rate of the reversible absorption or volatilization of the solutes into and out of the stationary phase. In high-performance liquid chromatography (HPLC), the mobile phase is a liquid that flows through the stationary phase and the separation of compounds is determined by the relative solubility of the solutes in the two phases. In ion-exchange chromatography (IEC), the mobile phase is usually water that flows through a charged stationary phase and separation is achieved for ionic solutes that bind with the stationary phase. Supercritical-fluid chromatography (SFC) uses CO_2 as the mobile-phase gas which, under pressure, is in its supercritical state, in which the CO_2 has characteristics of both a liquid and a gas. Both stationary phases for GC and HPLC have been employed for SFC.

Volatile flavour compounds are usually analysed by GC and those that are non-volatile by HPLC. Since only ionic species are separated by IEC, it is used to separate organic acids and salts. Volatile flavours found in fats and oils can be measured directly by SFC, as lipids are soluble in supercritical CO_2 and do not foul a column as they do in GC.

GC became commercially available in the mid-1960s. About a decade later, commercial HPLC and IEC instrumentation became available, followed by SFC in the following decade.

Gas chromatography. Since only volatile flavour compounds move in the mobile phase, any non-volatile compounds must be removed before injection, because columns are destroyed by lipids, carbohydrates, proteins and minerals found in most foods. Sample preparation may involve headspace analysis, distillation, preparative chromatography, solvent extraction or a combination of these techniques.[11] Care must be taken during sample preparation to ensure that flavour compounds are not formed or degraded (transformed). In most GC instruments, a liquid sample is injected onto a heating block that volatilizes the solutes into the mobile phase. Even if artefacts are not developed during sample preparation, they can be formed in the injection port. For example, it has been shown only recently that onion and garlic flavour compounds cannot withstand the normal temperatures in the injection port.[12]

Originally, GCs utilized packed columns for the stationary phase. However, in the early 1980s, capillary columns were developed that improved resolution and used less sample. Today, packed GC columns are used for preparative-scale separations, whereas capillary columns[13] are used for analyses. There are four main types of columns in use today: non-polar (OV-1, SE-30), intermediate polarity (OV-17, OV-225), polar (carbowax) and chiral (cyclodextrin). Stationary phase selection involves both intuition and knowledge of chemistry, and help with both can be obtained from the column manufacturer and the literature. However, selecting the best stationary phase is not as necessary as it once was, since the high efficiency of capillary columns often results in separations even when the phase selectivity is not optimal.

Numerous detectors are available for GC, each one offering advantages in either sensitivity or selectivity. The most common detectors are flame ionization (FID), thermal conductivity (TCD), electron capture (ECD), flame photometric (FPD), and photoionization (PID). Detectors with limited applications are nitrogen-phosphorus (NPD) and electron laser capture (ELCD). One can even use the human nose as the detector; a hand-made funnel, called a sniffer port, can be attached to the GC. The commercially available unit has been christened CHARM (combined hedonic and response measurements) analysis.

'Hyphenated' GC techniques are those that combine GC with another major technique; these include Fourier transform infrared (GC–FTIR), mass spectroscopy (GC–MS), and atomic emission spectroscopy (GC–AES). These 'hyphenated' techniques allow compound identification as well as quantification and are rapidly replacing wet chemical methods for identifying the flavour molecules.

High performance liquid chromatography. HPLC can be used to analyse for any compound that is soluble in the mobile phase. However, because it takes longer and the peaks are broader than GC, volatile compounds are not usually analysed by HPLC. Since more compounds are soluble in a liquid mobile phase, sample preparation is not as rigorous for HPLC. Many times, the food sample can simply be extracted with an organic solvent, diluted with the mobile phase, filtered and shot on the column. As long as a guard column, which can be changed often to remove irreversibly adsorbed material, is attached in front of the analytical column, destruction of the column is minimal. However, the more complex the sample matrix, the harder the chromatography becomes. Thus, it is common to perform a sample clean up using small disposable chromatographic columns prior to the HPLC analysis.[14]

There are two common types of HPLC columns: normal phase and reverse phase. In normal phase chromatography, the stationary phase is polar (silica gel) and the mobile phase is non-polar (organic solvents). In reversed-phase chromatography, the stationary phase is non-polar (C-18, C-8) and the mobile phase is polar [water, CH_3OH, CH_3CN, tetrahydrofuran (THF)]. There are also special columns available with intermediate polarity for rare problems (amine, diol, phenyl) and cyclodextrin for chiral separations.

The workhorse of HPLC detectors is the UV/VIS detector set at a specific wavelength. However, it must be remembered that each individual analyte will have different chromophores, and thus absorb different amounts of light at different wavelengths. This means that when developing a new HPLC analysis, it is necessary to isolate and purify enough of a flavour compound to determine the relative response of the detector compared to the standard (either internal or external) before quantitative estimates can be made.

A fluorescent detector is more sensitive than UV/VIS, but few flavour compounds possess the ability to emit electromagnetic radiation.

Considered the universal detector, refractive index (RI) measures all compounds that possess a different ability than the solvent to bend light. This type of detection yields both positive and negative peaks and is not as sensitive as most other methods of detection. This frail detector also suffers from temperature and flow-rate variations and cannot be used with gradient elution. But detection by RI is used for sugars, since they are non-volatile and do not contain a chromophore.

Another universal detector uses IR spectroscopy, but this technique is used more for the identification of compound classes than for quantitative analysis.

Electrochemical methods are also utilized for HPLC detection. Measuring the change in current as the analyte is oxidized or reduced, amperometric detectors are very selective and sensitive, especially when the current comes in pulses.

Ion exchange chromatography. This is actually a special case of HPLC, utilizing the same physical system of pumps and tubing. In this case, buffered water is used and the pH or ionic strength of the water can be varied for gradient elution. The column is usually charged positive or negative (SO_3^-; NR_3^+) and ions such as salts and acids are separated. A conductivity detector is used to measure the current that can flow through the mobile phase after it leaves the column to establish when ions are coming off the column.

Supercritical fluid chromatography. SFC is a relatively new technique that is a hybrid between GC and HPLC. This system can use either GC or HPLC columns and detectors. However, most systems use a GC oven for temperature control and an FID detector with either an HPLC column (packed) or a GC-type capillary column. The FID detector must be modified from those used in GC, as the CO_2 mobile phase acts as a fire extinguisher. Since the chromatography produces peaks that are broader than GC, but not as broad as HPLC, SFC is not used when a sample can be analysed by GC.

Non-volatile polar compounds are not soluble in supercritical CO_2, so they cannot be analysed by SFC and if injected into the system will foul the column. Small amounts of methanol can be added to the mobile phase to increase the solubility of polar substances in supercritical CO_2, but the methanol interferes with FID detection, so a UV/VIS detector is usually used. The best use of SFC using FID detection is for flavour compounds that do not contain a chromophore, but are in a lipid matrix.

Hyphenated techniques. As foods are so complex, with many compounds not separating well using just one technique, pre-purification is often necessary. Simply separating a complex mixture by GC, HPLC, or SFC may give overlapping peaks that cannot be resolved well. The column and/or mobile phase can then be changed, which changes the partitioning efficiencies, to remove interfering compounds.

Heart-cutting is when two GCs are hooked up together (GC–GC) and a portion of the eluent from the first GC is placed onto the second GC. This has usually been done with a packed column on the first GC and a capillary column on the second GC because of sample sizes.

HPLC–GC is a technique where a fraction collected from an HPLC containing volatile compounds is injected into the GC.

There are commercial units today that combine supercritical CO_2 extraction with automatic injection into a GC. This is not SFC–GC, because no chromatography is performed prior to injection onto the GC. Because most non-polar compounds and high molecular weight polymers, like proteins and complex carbohydrates, do not dissolve in supercritical CO_2, this is an efficient method to clean up a sample with minimum effort. However, problems can occur with fatty foods if pressures are too high; lipids are extracted and can foul the GC column.

The following hyphenated techniques allow compound identification as well as quantification and are rapidly replacing wet chemical methods for identifying the chemical structures of flavour molecules: GC–MS, GC–FTIR, and HPLC–MS.

SAMPLE HANDLING AND ARTEFACTS

Sample Selection

The first step in any analysis is to select the samples to be analysed, which may not be as straightforward as it seems. Sample selection depends first on what the problem is that the analysis is to solve. If one is trying to control the quality of a food product, then a representative sample is absolutely necessary; however, if one is trying to identify a flavour or off-flavour, then the strongest flavoured samples need to be selected. Using the most intense samples will increase the probability that relatively insensitive machines can pick out the compounds of interest. One must also think of the resources available. How accessible are the most ideal samples? Are they available only once, or do they occur seasonally or on a routine basis? How much is available? How perishable are they? What is the required turnaround time? Are samples going to be pooled or replicated? Do different portions need different analyses? For example, the surface versus the bulk of a prepared steak contains very different flavour compounds. One of the most important things to consider is how to monitor samples in order to detect and prevent contamination and abuse.

Although we can simply put a piece of bread or hot sauce into our mouths and obtain a sensory response, a desirable instrumental response occurs only after the flavour compounds have been extracted from the food, concentrated, and placed into the instrument.

Homogenization

Particle size reduction of a food is usually the first step in extracting flavour. For sensory analyses, the food is cut and chewed well. For instrumental analyses, this is generally accomplished by grinding (dry sample), cutting or slicing (whole foods), shearing or blending (wet samples) or shattering (seeds). Particle size distributions can be observed by sieving (dry samples) or microscopy (wet and dry samples).

Mixing

Only a portion of a sample is actually analysed instrumentally. If the food sample is not well mixed before that portion is removed, then the analysis may vary with each portion. Problems to overcome include *classification, stratification, phase separation* and *agglomeration*. In heterogeneous samples, such as foods, different particles contain different amounts of flavourants. For example, essential oils in citrus fruits are mainly in cells of the flavedo layer. If the particles differ in shape and size, they can separate during manipulations. This is especially true when the particles are also different in density, electrostatic charge or surface tension. The flavourants can be encapsulated in structures or entrapped by hydrophobic–hydrophilic interactions.

Sample Preparation

Sample preparation includes any operation performed on the test portion prior to analysis. Problems to guard against are moisture variations during weighing, incomplete dissolution of the attribute or sample during dilutions, and volume changes with temperature fluctuations. Also, since the methodology used for *'clean-up'* greatly influences a flavour profile, one needs to determine why the clean-up step is there, how it works and what it does to the sample.

Most fresh plant and animal tissues contain active *enzyme* systems that, once cell walls are disrupted, quickly alter the flavour profile. Thus, the rapid inactivation of enzymes during sample preparation is essential, but one must be aware of the artefacts or interferences contributed by means of enzyme inactivation. Commonly, thermal processes are used to inactivate enzymes. However, this may result in the loss of unstable or volatile flavour compounds. Homogenizing the food in methanol[15] avoids this problem, but it may interfere with isolation by decreasing the polarity of the aqueous food slurry and forming methyl ester artefacts. Enzymes can also be inactivated with the use of heavy metal salts containing Cu^{2+}, though this metallic ion can catalyse lipid oxidation. So far, no obvious disadvantage to the isolation of flavour compounds has been found when foods containing enzymes are treated with high concentrations of neutral salts, such as an equal volume of saturated $CaCl_2$ solution, to inactivate them.

Sometimes enzymes are added or activated to help in the analysis of food samples. Carbohydrases and proteases are often added to digest the foods, so the compounds of interest are more easily freed from the sample. However, during this digestion, many of the flavour compounds can also be transformed. Thus, digestion is not recommended for flavour analysis.

Long isolation procedures may permit fermentation to occur. Antimicrobial agents are sometimes added to control this.

Besides enzymatic and microbially induced changes in flavour profile, chemical changes can also occur. It is common practice to add antioxidants or prepare the sample under N_2 or CO_2 to avoid oxidation. High temperatures ($>60°C$) for extended periods can promote Maillard browning reactions. Thus, reduced temperatures during sample preparation are recommended.

Since flavours occur at such low concentrations, contamination of the test portion before analysis is a real problem. Most food is now packaged and handled with plastic containers; small amounts of plasticizers can migrate into the foods and be detected by GC–MS.

Isolation and Concentration

Direct analysis of the *headspace* vapours above a food product is the ideal method to isolate flavours, but the primary problem is the low concentration of flavour compounds in the headspace. Headspace analysis has found substantial application in flavour studies where trace analysis is not necessary.

Purge and trap. The equilibrium headspace vapours above a solid food or the food itself may be purged with an inert gas in order to obtain a large volume of headspace gas for analysis. The flavour compounds are then concentrated using cryogenic traps or adsorption columns. The problem with cryogenic traps is that the most abundant volatile in foods is water, so an additional step is generally needed to extract the flavour constituents from the water. Volatile flavours are often trapped on adsorbent materials to avoid this co-condensation of water. If the adsorbent has a minimal affinity for water, this eliminates the need for solvent extraction and the associated problems. Typical adsorbents used are charcoal (activated carbon), Porapak Q (polymer of ethyl vinyl–divinyl benzene), Tenax GC (polymer of diphenyl–phenylene oxide), XAD resins (polymers of divinyl styrene, acrylic ester or sulfoxide).

Distillation. Distillation is a rather broad term that includes any technique that vaporizes a liquid mixture through the addition of heat and subsequently collects the vapours by removal of the heat. It takes advantage of the difference between the volatility of flavour components and the non-volatility of the major food constituents. A simple distillation is not usually used in flavour analysis because of the small concentration of flavour components and their delicate nature. Rather, extra water is added to the food sample to lower the boiling points of the flavour compounds and help carry them out into the collection vessel. This technique, called hydrodistillation, utilizes co-distillation of water to transfer the flavour volatiles into the distillation head. Steam distillation is when the water is added to the food sample continuously in the form of steam. The distillate is a very dilute solution of volatile flavours and water which must be extracted with an organic solvent, and the solvent dried (*e.g.* anhydrous $MgSO_4$) and concen-

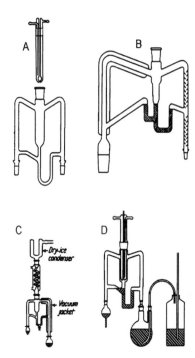

Figure 4.4   *Various Likens and Nickerson devices for the isolation of volatiles by simultaneous steam distillation extraction. A: Original apparatus. B: Modification includes a solvent arm with Vigreux indentations and an insolated solvent arm, suitable for extraction with 'heavy' solvents. C: Modification with vacuum jacket to minimize premature condensation and dry-ice condensor to reduce volatilization losses. D: Modification where steam is generated in a separate vessel. From Maarse and Belz.[16]*

trated before analysis can be performed. The distillation can be performed directly into a solvent trap to help speed the extraction step.

In 1964, Likens and Nickerson developed a method that combines both the distillation and extraction steps, called simultaneous distillation/extraction (SDE). As shown in Figure 4.4, distilling the extracting solvent during the collection of the primary aqueous distillate allows for a very efficient flavour extraction. However, this extended heat treatment may be deleterious to some of the unstable flavour compounds.

Vacuum distillation is another way to lower the boiling points of flavour compounds; the distillation is performed under reduced pressure. When a very high vacuum is used, this is called molecular distillation. Usually utilized in high fat and/or low particulate samples, the vacuum requirements limit this method to food samples containing essentially no water. For molecules to volatilize out of a liquid food matrix efficiently, they must be spread out in a thin film to increase the surface area of the liquid–gas interface, so very short distances must be provided between the condenser surface and the thin film of food. This can be accomplished with two very different designs: (1) falling film and (2) spinning disk. In falling film, the lipid-rich sample is dripped onto the inside surface of a pipe heated from the outside by rotating wipers that keep the food matrix in a thin film. As the film falls, volatile molecules escape and are collected on a condenser that is found in the centre of the pipe. With spinning disk, the lipid-rich sample is dripped onto the centre of a tilted rotating circle that is in close proximity to a flat circular surface cooled with water (condenser). The speed of the rotation and the viscosity of the sample control how thin the film will be on the surface of the disk. As the sample migrates to the outside of the circle, flavour molecules escape to be collected off the flat condenser. A significant advantage of molecular distillation is the decreased opportunity for artefact formation or contamination.

Extraction. Most flavour compounds exhibit substantial solubility in organic solvents, but unfortunately lipids are also soluble in organic solvents. Therefore, in high-fat foods, the solvent extract (fats and flavours) may need to be treated further by steam distillation, molecular distillation, chromatography, dialysis or liquid CO_2 to separate the flavours from the fats. Thus, solvent extraction is best performed on fat-free foods, such as alcoholic beverages, fat-free bakery products, fruit or berry juices and some vegetables. If the food sample is dry, simple percolation through a column works well. To reduce solvent use, it can be recycled via distillation through a Soxhlet apparatus. For liquid food samples, the simplest method is batch extraction using separatory funnels.

The major problem with batch solvent extraction is emulsion formation in the separatory funnel. Emulsions may be broken by the use of centrifugation, addition of a saturated NaCl solution, slow stirring or pH adjustment. Bath solvent extraction is also tedious and labour intensive. Continuous liquid/liquid extraction generally provides more efficient extraction and reduced labour requirements, but at a greater cost in equipment and time.

When CO_2 is liquified under pressure, it is a useful non-polar solvent. Liquid CO_2 is used to separate volatile flavour compounds from high-fat foods, as it does not extract lipids very well. Liquid CO_2 has also been used to separate the polar and non-polar constituents of orange oil.

In the presence of water, liquid CO_2 contains a small amount of carbonic acid: $H_2O + CO_2 \rightarrow H_2CO_3$; thus its non-polar solvent properties are modified to provide some polar character. As pressure is increased, more carbonic acid is formed. Above a pressure of 73 bar and temperature of $31\,°C$, CO_2 becomes supercritical, in which state it becomes a better solvent, holding a much higher concentration of compounds. The solute load of CO_2 can be controlled; the lower the temperature and the higher the pressure, the more soluble polar compounds become. Supercritical extraction (SCE) units fit on GC instruments which perform extraction and analysis of solid food samples automatically.

Dialysis. This technique is good for small molecules in fat samples, such as in the isolation of cheddar cheese flavour. A membrane is used to hold back the large triglycerides and allow the small flavour molecules to pass. However, recovery of flavour molecules decreases rapidly as the molecular size of the sample increases.

Derivatization. Chemically modifying flavour compounds offers several advantages. Usually the derivative imparts UV-absorbing properties, which thus aids in HPLC analysis using UV detection. Many unstable and highly volatile flavour compounds become stable and non-volatile upon derivatization. Because derivatives are specific to a functional group, this greatly simplifies the problems of separation and detection.

Concentration. Most of the extraction and isolation methods produce a dilute solution of volatiles that needs to be concentrated for analysis. Evaporation is typically used for flavour isolates in organic solvents, and takes advantage of the differences in boiling points between the flavour compounds and solvent. A disadvantage of evaporative techniques is that flavour compounds may be lost by co-distillation. Since flavour isolates typically contain water from the food product, care must be taken to remove water prior to concentration to prevent steam distillation and substantial flavour loss.

Flavours can also be adsorbed from dilute aqueous solutions by charcoal, silica gel, alumina, porous polymers, and other adsorbents. Typically, an aqueous distillate is passed through a column of adsorbent, the adsorbent is rinsed with water, and then eluted with an organic solvent. These same adsorbents are used to concentrate the dilute flavour volatiles in a gas (*e.g.* headspace).

DATA HANDLING

Design of Experiments

When should the various tests be used? This depends upon the objective of the experiments. If quality control is being performed on a product that should not change, then a simple triangle test may suffice. If a new product is being developed and the origin of an off-flavour needs to be known, then both sensory and instrumental tests are needed. Sometimes instrumental analysis of a non-flavoured compound in a food can be used to assess the status of a flavour; for example, both pentane and hexanal have been used as indices of the oxidative flavour deterioration of fats in foods. However, the essentially odourless pentane seems to provide a better index of the actual flavour intensity than does hexanal, which definitely contributes to the flavour. So the best advice is to determine objectives before performing any type of analysis for flavours.

There has been a consistent desire of analytical chemists to equate 'subjective' with sensory and 'objective' with instrumental analysis. However, if the instrumental method is not truly measuring the total flavour, then it may not be helpful to solve the flavour objective of one's experiment. Sensory data have become much more 'objective' as we find ways to permit people to respond in a consistent and predictable manner. A panel of selected and trained people can consistently and quantitatively measure differences that we cannot measure instrumentally.

Sensory and instrumental analyses may be correlated in one of two ways. The instrumental information may merely reflect the status of the flavour, such as in the example above of using pentane to determine the extent of lipid oxidation, or it may be analysed to identify the actual compounds that contribute to the flavour.

Chemometrics

Flavour changes frequently result from subtle alterations in many components, rather than drastic changes in a few, so analysis of one variable at a time is not normally sufficient. Chemometrics uses multivariate statistics to determine the correlation between sensory and instrumental analyses. Treating two or more variables simultaneously requires the use of matrix algebra, vectors, eigenvectors and eigenvalues. Techniques such as nearest-neighbour analysis can serve to classify unknown samples if there are samples available from each known category. Techniques such as clique analysis and multidimensional scaling can suggest categories when researchers do not know beforehand what categories exist. Factor analysis can provide clues to unknown underlying variables. Reducing the number of variables is often accomplished by principal-components analysis, which reduces the variability in a limited number of variables, and discriminant analysis, which focuses on combinations of variables that distinguish among categories.

Artificial neural networks can be used to train a computer to recognize a pattern, evaluate a new unknown sample, and place it within one of previously determined categories (Figure 4.5). As computers increase in speed and memory, artificial neural networks can be connected to sensor arrays to mimic the human nose.

Electronic nose. A commercial product recently on the market, AromaScan, emulates the human nose. Developed predominantly at the University of Manchester Institute of Science and Technology's Department of Instrumentation and Analytical Science, it is based on an array of sensors that are

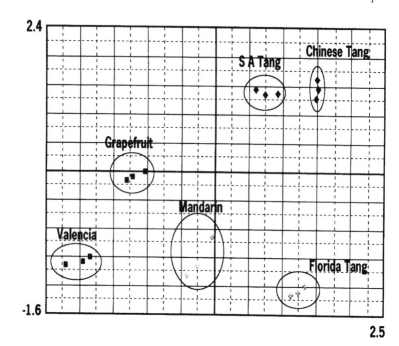

Figure 4.5   *A two-dimensional map of six different citrus oils with different aroma characteristics produced by computer manipulation of the data from an array of sensors made of electrically conducting organic polymers. Tang = tangerine. Reproduced with permission from AromaScan.*

formed from electrically conducting organic polymers.[17] The sensors function by the adsorption of volatile chemicals onto the polymer surface causing a change in electrical resistance. The extent of this change depends on the chemical compositions of the aroma and the physical and chemical structure of the polymer. The sensor array produces signals which are digitized and processed via computer to produce a distinctive and unique pattern of response for the sampled aroma (Figure 4.6). When combined with a neural network, this instrument can be taught to recognize the pattern of an aroma of a product. Because both adsorption and desorption of volatiles at the polymer surfaces are quick and all the sensors operate simultaneously, this can be used in quality control or quality assurance programmes as the first line of defence for routine analysis. Thus, only the samples with an identified divergence from acceptable aroma variations need to be analysed further.

Correlations with instrumental and sensory techniques.   There is no direct correlation of this new instrumental technique with current instrumental methods. Since a single compound produces a signal in many sensors and a single sensor has

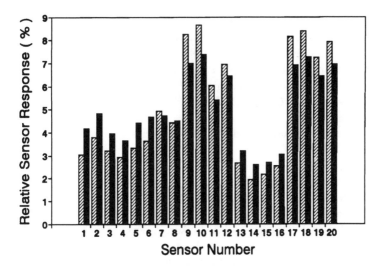

Figure 4.6 *Comparison of patterns obtained from an array of 20 sensors (made of electrically conducting organic polymers) of corn samples evaluated as good (solid bars) and bad (hatched bars) by odour panel. From Persaud, et al.*[17]

a broad sensitivity to different compounds, the identification of specific volatile compounds still needs to be performed via chromatographic methods.

As this instrument is emulating the human nose, it correlates fairly well with sensory panels doing difference testing when the neural network is trained with many samples that contain the variations found acceptable and unacceptable to the sensory panels. As the sensitivity of sensors to specific compounds may be larger or smaller than that of the human nose, there are cases where complete correlations will be impossible using the current commercially available sensor array. In these cases, the conducting polymers on the sensor array may need to be custom developed for such specific applications.

REFERENCES

1. S.S. Stevens, in *Sensory Communications*, ed. W.A. Rosenblith, MIT Press, Cambridge, Mass., 1961.
2. B. Wenzel, *Ann. Rev. Psychol.*, 1954, **5**, 111.
3. J.F. Caul, *Adv. Food Res.*, 1957, **7**, 1.
4. H. Stone, J. Sidel, S. Oliver, A. Woolsey and R.C. Singleton, *Food Technol.*, 1974, **28**, 24.
5. M. Meilgaard, G.V. Civille and B.T. Carr, *Sensory Evaluation Techniques*,

CRC Press, Boca Raton, 1987, vols. I and II.

6. M.H. Gillette. In *Source Book of Flavours*, ed. G. Reineccius, 2nd edn, Chapman & Hall, New York, 1994.

7. N. Larson-Powers and R.M. Pangborn, *J. Food Sci.*, 1978, **43**, 41.

8. W.E. Lee III and R.M. Pangborn, *Food Technol.*, 1986, **40**, 71.

9. *Introduction to the Chemical Analysis of Foods*, ed. S.S. Nielsen, Jones and Bartlett, Boston, 1994.

10. *Source Book of Flavors*, 2nd edn, ed. G. Reineccius, Chapman & Hall, New York, 1994.

11. M.J. Lichon, *J. Chromatogr.*, 1992, **624**, 3.

12. E. Block, D. Putman and S.-H. Zhao, *J. Agric. Food Chem.*, 1992, **40**, 2431

13. A. Mosandl, *J. Chromatogr.*, 1992, **624**, 267.

14. C. Markel, D.F. Hagen and V.A. Burnelle, *LC–GC*, 1991, **9**, 332.

15. P. Schreier, F. Drawert and A. Junker, *J. Agric. Food Chem.*, 1976, **24**, 331.

16. H. Maarse and R. Belz, *Isolation, Separation and Identification of Volatile Compounds in Aroma Research*, D. Reidel, Dordrecht, 1985.

17. K.C. Persaud, A.A. Qutob, P. Travers, A.M. Pisanelli and S. Szyzko. In *Olfaction and Taste XI*, ed. K. Kurihara, N. Suzuki and H. Ogawa, Springer-Verlag, Tokyo, 1994, p. 708; A.M. Pisanelli, A.A. Qutob, P. Travers, S. Szyszko and K.C. Persaud, *Life Chem. Rep.*, 1994, **11**, 303.

Chapter 5

Teaching Flavour Concepts

The difficulties outlined in Chapter 1 show that both research into and teaching about flavours is a challenge. We believe that a holistic approach to food flavours is critical for anyone working in the field. But how does one teach a holistic approach? No single person can be expected to know everything about the food systems that flavours are incorporated into. We believe that what needs to be taught is how to define and, then, find the information when it is needed. Thus, this chapter is devoted to showing some flavour applications as case studies or problems that can be utilized to develop critical thinking skills. It also provides a framework in which the knowledge described in the previous chapters (and from other resources) can be placed. Obviously, there will be some non-flavour issues brought out in these case studies, but in a holistic approach we cannot limit the students too much if we wish to encourage their thinking skills.

A brief explanation of the teaching methodology, and some application problems, and their solutions are given below.

PROBLEM BASED LEARNING

An instructional method that is characterized by the use of 'real world' problems as a context for students to learn critical thinking and problem solving skills has been labelled 'problem based learning'. While acquiring the knowledge of the essential concepts of a course, students also acquire life-long learning skills which include the ability to find and use appropriate learning resources. The process involves problem presentation, learning issues, prioritization, peer teaching and reiteration.

Problem Presentation

Students are presented with a problem (case, research paper, videotape, etc.). Students (in groups) organize their ideas and previous knowledge related to the problem and attempt to define its broad nature.

Learning Issues

Throughout the discussion, students pose questions, called learning issues, on aspects of the problem that they do not understand. These learning issues are recorded by the group. Students are continually encouraged to define what they know and, more importantly, what they do not know.

Prioritization

Students rank, in order of importance, the learning issues generated in the session. They decide which questions will be followed up by the whole group, and which issues can be assigned to individuals, who later teach the rest of the group what they have found. Students and instructor also discuss what resources will be needed in order to research the learning issues, and where they could be found.

Peer Teaching and/or Reiteration

When students reconvene, they explore the previous learning issues, integrating their new knowledge into the context of the problem. Students are also encouraged to summarize their knowledge and connect new concepts to old ones. They continue to define new learning issues as they progress through the problem. Students soon see that learning is an ongoing process throughout their careers and lives, and that there will always be (even for the teacher) learning issues to be explored.

TONGUE AND NOSE

Problem

A food technologist in a snack food company was asked to reformulate their major product line, Cruncheeze, to produce a low sodium, low calorie version. However, on his first few attempts, he found that the reformulated product was bland with a bitter, sour after-taste. Knowing of your interest in taste physiology, he turns to you for help.

Prior knowledge

(1) Lipids (fats) as well as carbohydrates (sugars) are high in calories; thus both fat substitutes and artificial sweeteners were probably used in the reformulation.

(2) Low sodium requires that salt was not used, so sodium substitutes were probably added.

(3) Cruncheeze sounds like a cracker or extruded product that contains cheese or cheese flavourings.

Learning issues

(1) What makes foods taste bitter and sour? *[Both sweet and bitter perception rely on the specific flavour compound binding with a receptor site. However, for the perception of sour and salty, channels in the taste cells are either blocked by the proton to constrain potassium ions (K^+) to the inside of the cell (sour), or permeable to sodium ions (Na^+) to allow them into the cell (salty). See Chapter 3.]*

(2) How do taste receptors work? *(This is another way of asking Item 1. See Chapter 3.)*

(3) What kinds of compounds have bitter or sour flavours? Where might these originate in the reformulated Cruncheeze? *(Bitterness can be expressed by a wide range of molecules, with varying sizes and functional groups: aliphatic or aromatic compounds, straight chained or polycyclic, glycosides or aglycones, but they are usually non-volatile;[1] acids are sour. These bitter compounds may always have been in the food, but in the presence of fat they were not perceived.)*

(4) How do the major taste sensations interact? *(Some interaction occurs among sweet, salty, sour and bitter. In addition, there is interaction of the taste sensations with olfaction.[2])*

(5) What components of food besides salt (NaCl) contain sodium? *(Many organic acids are added to foods as their sodium or potassium salt; also, many flavourants that are added as powders are dispersed on salt.)*

(6) What are artificial sweeteners, fat replacers and salt substitutes made of? How do they work? *(Artificial sweeteners are synthetic compounds sold to replace sugar, examples being saccharin, cyclamate, acesulfame-K and aspartame. Fat replacement systems have been created using gums, starches, dextrins and proteins. The common element is water and the key to success is how the water is controlled to make it provide the functionality of the missing fat. Salt substitutes usually contain some herb and/or spice flavours or flavour enhancers to provide flavour appeal; there is no replacement for Na^+ that gives a purely salty taste. KCl is both bitter and salty.)*

Problem

In the meantime, the food technologist heard about the fruit of the miracle plant that has no flavour on its own, but turns a sour lemon very sweet. He asks you to find out about its active ingredient, miraculin (which is a protein) and whether he could use it in his product.

Prior knowledge

(1) A lemon is sour because citric acid reduces the pH.
(2) At a low pH, some proteins are denatured and precipitate, for example, milk proteins in the making of cheese.

Learning issues

(1) Can proteins be sweet? What other compound classes besides sugars are sweet? *[Simple carbohydrates (monosaccharides and disaccharides); D- and L- amino acids (most D-amino acids are sweet, only some L-amino acids are sweet); high impact sweeteners (saccharin, aspartame, etc.); plant proteins (monellin and thaumatin); chloroform; simple salts of beryllium and lead.]*

(2) Are there any chemicals other than miraculin that change their flavours in the presence of other compounds? Or by their presence change the flavour of other compounds? *[Although the protein curculin has a sweet taste by itself, its sweetness is greatly enhanced in the presence of acid or in the absence of divalent cations. Vegetable- and meaty-taste enhancers include monosodium glutamate (MSG) and inosine 5'-monophosphate (IMP).]*

(3) Enzymes are proteins; could miraculin be an enzyme? *(No, it is not.)*

(4) What could be miraculin's mechanism of action? *(Miraculin was isolated from the berry of a shrub* (Richardella dulcifica) *in 1969.[3] A suggested mechanism of action is that miraculin binds to the sweet receptor, inducing a conformational change in the receptor that permits its full activation only at low pH. Acids offer that low pH and hence elicit a sensation of sweetness.[4])*

Problem

After you explain how the tongue works to the technologist, he realizes that sweetness and bitterness are somewhat related. However, he is not sure how all this information on taste physiology can help him. He was very thankful for the list of artificial sweeteners, fat replacers and salt substitutes that you provided him. He thought he would go back to the laboratory and try some of them.

The technologist has hay-fever and, during the fall months, he has a chronic stuffy nose. In September, he found a low fat variation that he thought would work in his product. Since he thought it was comparable to the regular product, he sent it off for taste testing. A month later, he was given the results: his variation looked good and was sweet enough, but it was rejected because it did not smell cheesy enough!

He has dinner at your house and complains about all his hard work going for naught! Since you did such a good job explaining how the taste system works, he asks you to explain olfactory function.

Prior knowledge

(1) A stuffy nose reduces the ability to smell.
(2) Flavours change with time. If it took a month for the taste test to be done, then the data may be irrelevant.

Learning issues

(1) How does olfaction function? *[Volatile molecules enter the olfactory system either through the anterior nares of the nose (nostrils) or the posterior nares in the roof of the mouth. Once these compounds reach the olfactory epithelium at the top of the nasal passages, they are collected by the mucus and carried by cilia to the receptor cells. These receptor cells extend through the cribriform plate, a thin bone at the base of the skull, to contact the second-order cells in the olfactory bulb. It is these cells that transmit olfactory signals to numerous regions in the brain;[5] see chapter 3.]*

(2) Are there separate categories in smell, like sweet and sour in taste physiology? *[There have been some attempts to classify odours into primary categories,[6] however, this is now considered futile because: (1) no underlying physical continuum along which odours vary systematically has been established, (2) odour naming is idiosyncratic to the individual, (3) the hedonic component of aromas is difficult to eliminate and (4) the number of classifications found depends on the number and quality of odours studied.[5]]*

(3) How do individuals differ in their ability to smell and taste flavours? *(There is great variability in chemical sensitivity, even in the normal population. Perception of chemicals also declines with age. There are also genetic factors[7] and aspects of health and reproductive status that affect sensitivity.)*

(4) What are the different types of taste tests? Which type should have been used in this case? *(See Chapter 4; a triangle or duo–trio test is sufficient.)*

(5) What chemicals cause cheese flavour? *(Flavour compounds in cheese come from the enzymatic or microbial liberation of volatile fatty acids from milk fat and the degradation of casein into low molecular weight peptides and amino acids. The specific compounds are unique to the microbe or enzyme and determine the variety of cheese. Utilizing strong flavours at least 20 times more concentrated than in natural cheese, the technologist could add enzyme modified cheese, butter or cream. These are obtained by controlled proteolytic or lipolytic enzyme treatment[8] of a previously manufactured traditional natural cheese or other dairy products.)*

ONION

Problem

Imagine that you currently work as a food technologist in a local company that processes regional vegetables for a limited market. Your president has decided that it is now time to expand the small company and extend the markets into a region that is ten times larger than at present. Your lightly pickled green beans have been selling very well on the market because of a special proprietary process that ensures a fresher taste. However, the president knows that there is a problem with the green beans: the shelf-life of the fresh flavour of the green bean product is only three months; at that time the product takes on a metallic taste. This does not matter in the limited market because the turnover in the stores is quick. However, with expansion and therefore less control over the stores that sell the product, he expects that the product may not be sold and eaten within the three month period.

Your president tells you to work on increasing the shelf-life of these lightly pickled green beans. His proprietary process has been simply to add onion oil at a subliminal concentration to the pickling brine used after blanching, but before retorting in enamelled tin-plate cans. Beans processed without this 'magic' ingredient do not have as good a taste initially, but do not go bad within three months.

Prior knowledge

(1) Pickled means that the green beans have been acidified, probably with vinegar (acetic acid).
(2) Subliminal means below threshold or that the onion oil cannot be smelled.
(3) Blanching means to boil the vegetables for a short time to denature degradative enzymes.
(4) Retorting means to sterilize foods in containers under pressure. A retort is a large strong vessel capable of withstanding steam pressures of 103–207 kPa (15–30 psi).

Learning issues

(1) What does 'fresher flavour' really mean? *(This is a subjective question and leads to a discussion of the transformation of sulfur compounds in an onion. Given that onion flavour is developed enzymatically, fresh could mean being lachrymogenic, or containing the thiosulfinates, see Chapter 2. This also leads to a discussion of the subjectiveness of taste tests and how important it is to define your goal via a control sample or something tangible and to make sure that everyone on your team is in agreement with your definition.)*
(2) Is shelf-life really the time a product is on the shelf in the store? *(It is actually defined as the length of time before quality parameters begin to diminish.)*
(3) How are green beans pickled? What is the process? *(Beans are cleaned and*

cut, then blanched to degrade enzymes. The blanched beans are put in cans and covered with a brine containing vinegar and salt, then canned under pressure.)

(4) What is onion oil? How is it made? What are its chemical components? *(Onion oil is the non-water-soluble fraction from the steam distillation of macerated onions and contains mainly dipropyl disulfide, with a flavour detection threshold of 3.2 ppb.[9])*

(5) What compounds produce green bean flavour? *[This should be considered of low importance for this problem. However, (Z)hex-3-enol, α-terpineol, and linalool are thought to impart the characteristic flavour to some varieties of canned beans.[10]]*

(6) Could the metallic flavour be from the cans? *(Green beans are normally canned in enamelled tin-plate cans because they are very corrosive; steel cans become thin over time, so only if defects are in the enamelled tin-plate cans, will they be the cause. This is a good question to make students think about metal flavours - metals in the uncharged state cannot be transported to the taste receptors; only charged metal ions can reach the receptor sites.)*

Problem

You decide the problem is with the onion oil and try to replace it with products that have an onion flavour in them. However, you do not have six months to wait to establish the answer, so you set up an accelerated shelf-life study. You discover that the metallic taste can be generated in the green beans within a week if the product is put into an oven at 35°C (95°F). So you set up an experiment in the laboratory using a bench-top hot water canner with five different samples: (1) no onion, (2) onion oil, (3) fresh onions, (4) dehydrated onions, and (5) onion paste. Then you put the samples into the 35°C (95°F) oven. When you come back in a week, a couple of the cans have burst!

Prior knowledge

(1) A hot water canner does not become hot enough to can vegetables unless they are below pH 4.0; a pressure canner is necessary to superheat the steam for food that is not acidic enough.

(2) If the temperature is elevated only slightly, the formation of the metallic off-flavour is accelerated.

Learning issues

(1) What went wrong? *(Some microbes can survive hot water canning and grow to give off gases.* Clostridium botulinum *is the most dangerous, but does not give off*

gases. Or possibly the metal in the cans was dissolved in the acidic brine and holes appeared.)

(2) What is the difference between how the different onion products are processed? (Onion oil is steam distilled, hence wet heat is used. Dehydrated onions are minced with water removed in a tunnel drier or freeze drier; hence dry heat is used with temperatures below the boiling point of water. Onion paste is onions macerated in vegetable oil preserved with citric acid and with salt and pectin or other thickeners added.)

(3) What is the difference in flavour compounds in these different onion products? (Onion oil is mainly disulfides. Dehydrated onions contain both the cysteine sulfoxide flavour precursor and the thiosulfinates and thiosulfonates, while any disulfides would have been lost with removal of the water. Onion paste also contains cysteine sulfoxides, thiosulfinates and thiosulfonates, but the disulfides would also be present.)

Problem

You redid the experiment, but this time made sure that you had better hygiene in the lab and used a pressure canner for the recommended time and pressure. The canner was large enough to do the experiment in duplicate, so you left one set on the lab bench to simulate normal storage and the other set went into the 35°C (95°F) oven. After a week's time, you opened all of the cans in the oven and tasted them. The results were as follows:

(1) No onion (control) – Acceptable flavour.
(2) Onion oil – metallic flavour evident.
(3) Fresh onions – no metallic flavour, fresher flavour than the control.
(4) Dehydrated onions – no metallic flavour, but off-flavour evident.
(5) Onion paste – strong metallic flavour.

So you went back to your boss and told him that green beans could be made with fresh onions to do the job. The president was livid: he already knew that! How do you think the proprietary process had been discovered in the first place! You were told in no uncertain terms that if onions were found in the green bean cans, others would discover their secret ingredient.

So you decide to purée the onions in a gallon-sized metallic blender with a little water to make the pieces as small as possible. But when you open the lid, you are blinded with tears, so you close it up again. It was nearly lunchtime, so you let the closed blender sit for two hours before you opened it again. This time, no tears! You use half of it for today's experiment and put the rest into the refrigerator (4°C) overnight. In the morning, the onion purée had turned pink!

Perhaps fresh onions will not do the job. You knew that the plant manager would not accept using something that could not be used for two hours after preparation and then could not be left to sit overnight before it went bad.

You thought perhaps the manager might accept the purée if it were made as part of the brine solution, which is usually made up at the beginning of a shift and not used for an hour while the beans are being prepared and blanched. It is also used in total within that shift. So you try one more experiment, making the purée in the presence of the pickling brine. This time you receive a whiff of the lachrymator without even taking the cover off. After two hours, the amount of lachrymator was still worse than in freshly puréed onions!

Prior knowledge

(1) Onion oil is a thin liquid, dehydrated onions are a powder and onion paste is a thick viscous fluid.
(2) Pickling brine contains both acid and salt.

Learning issues

(1) For each process of making onion flavours, which compounds are responsible for metallic flavour, fresh flavour and off-flavour? *(Onion oil in high concentrations can have a metallic taste.[11] Fresh flavour is probably due to the presence of thiosulfinates. Without further information, the off-flavour compounds cannot be determined.)*
(2) What compound in onions makes your eyes tearful? (syn-*propanethial* S-*oxide.)*
(3) Why does the lachrymator disappear within two hours when the onion is puréed in water, but is stronger and lasts longer when the onion is in pickling brine? *(Propanethial S-oxide is stabilized by the formation of a sulfenic anion in the presence of acid and salt.)*
(4) Why did the onions turn pink overnight? *(Colourless anthocyanins under acidic conditions will give rise to colourful anthocyanin complexes with iron or tin[12] that could have come from the metal blender. Garlic similarly turns green. Although the discolouration of allium products poses serious problems to the industry, the nature and composition of the pigments responsible are not well understood.[11])*

Problem

You gave up on the shelf-life project and solved some other problems for the company. After three months, your boss asks if you have solved the lightly pickled green bean problem. You have not! However, you still have the beans on the lab bench, so you open them up and both you and your boss taste them:

(1) No onion (control) – Acceptable flavour.
(2) Onion oil – metallic flavour evident.
(3) Fresh onions – no metallic flavour, fresher flavour than the control.
(4) Dehydrated onions – no metallic flavour, fresher flavour than the control.
(5) Onion paste – strong metallic flavour.

Your boss was impressed that you had found the solution! Dehydrated onion powder would be an easy switch to make in the plant.

Learning issues

(1) What is the difference between the accelerated shelf-life study and this one? *(Temperature! At higher temperatures, rates of reactions usually go faster, however, new reactions can also happen at higher temperatures because the system is now above the activation energy of the new reactions.)*
(2) What causes the metallic flavour? *[Onion oil in high concentrations can have a metallic taste,[11] but this is not the only possibility. Sulfur compounds do bind tightly to metals, so the metallic taste could be due to a sulfur–metal complex with the metals picked up from the can. Some cans have a lacquer (epoxy or phenolic) lining that can degrade under acidic conditions.[13]]*

BEVERAGE

Problem

Imagine that you are a food technologist in a beverage company and you have been asked to reformulate the Zestaid product because some of the ingredients currently used are forecast to be in limited supply. Your company is planning to launch an aggressive marketing strategy and does not want to be caught with too little product to sell. Your 'new and improved' liquid beverage is to be as close as you can come to the current product. You were also told that only natural ingredients should be used.

The first thing you find is that the only ingredients that could be in limited supply are the natural extracts (except for the 10-fold orange oil) used to flavour the beverage.

Prior knowledge

(1) Natural extracts used for flavouring are the 'liquid' part of spices and herbs.

(2) Consumers will pay more for all natural products, because they think they are 'healthier'.

Learning issues

(1) What does 'fold' mean? *(Fold is a suffix meaning 'of so many parts' and denotes multiplication by the number indicated by the stem or word to which the suffix is attached, e.g. two-fold means twice as concentrated, 10-fold is ten times as concentrated. Essential oils are usually concentrated by removing non-flavoured components, such as simple hydrocarbons, e.g. limonene removal from orange oil.)*

(2) What are natural extracts? *[Essential oils are the volatile materials present in spices and herbs usually obtained by distillation. Oleoresins are the solvent extractables of ground spices and herbs after solvent removal. The common form of both essential oils and oleoresins is an oil-dispersible liquid. As most foods contain a high concentration of water, homogeneous mixing of small quantities of flavourful oils is difficult. Thus, other commercial products have been developed: dispersed products, usually called dry-soluble seasonings, are prepared by dispersing the essential oils and oleoresins on an edible carrier of salt, dextrose or flour; water-dispersible products are blends of essential oils and oleoresins with polysorbate esters or emulsifiers to make the oils disperse quickly into water; encapsulation (spray-drying or coating dispersed products) decreases degradation or evaporation of flavours and increases shelf-life of dry products.]*

(3) What chemicals are in essential oils and oleoresins? *[Not all of the chemical components of essential oils and oleoresins are known, but the major flavouring ingredients of most spices and herbs have been determined (see Chapter 2). The major non-flavoured chemicals of essential oils are non-oxygenated terpenes and of oleoresins are lipids.]*

(4) Where do you find possible substitutes for the natural flavours? *[The search needs to be limited to sources of extracts since you were instructed to use all natural ingredients. But this is not an easy task; at the 1995 Institute of Food Technologists National Meeting there were 133 companies under the combined heading of flavours, essential oils, and aromatic chemicals. Members of the American Spice Trade Association (ASTA) would also be a good place to start.]*

Problem

You go down to the plant and ask many questions about the current product. This particular beverage contains 10% juice. The juice is mainly single-strength orange 'pump-out', the rest is pineapple juice. Flavouring materials are also added back at 0.5%. After asking many people what orange 'pump-out' was, you finally decide that it must be orange juice without essential oils.

The flavouring materials used are 10-fold orange oil purchased from many different suppliers and a liquid flavouring mix containing extracts of lemon, orange, vanilla, ginger, cloves, cassia, nutmeg, and mace that is a proprietary blend produced by a single supplier, Biospherex. It is this flavouring mix that is limited.

You found that there are many individual extracts as well as blends on the market, but your company has specified the flavouring mix to have a certain refractive index, optical rotation, percent volatile oil and GC fingerprint, which only this single supplier, Biospherex, can generate.

Prior knowledge

(1) Refractive index is an intrinsic property of a substance defined as the ratio of the sine of the angle between the incident ray and the line perpendicular to the surface of the prism to the sine of the angle between the refracted ray and the same perpendicular line. It is usually used to measure the soluble solids in a product.

(2) Optical activity is the rotation of the plane of polarized light as it passes through a chiral substance, either in its pure state or in a solution.

(3) A GC fingerprint is a gas chromatogram printout of the entire profile of all the volatile compounds present in the sample.

(4) A percent volatile oil is simply a determination by weight of how much essential oil can be steam distilled out of a spice or spice extract.

(5) Besides 10% juice and flavouring, the rest of the beverage is probably water, sugar and citric acid.

Learning issues

(1) What is orange 'pump-out'? Is your thinking correct? *(Yes. This was the instructor's question which led to the next learning issue.)*

(2) Are there many different grades of orange juice on the market? How will they affect the flavour of the final beverage product? *(There is a 100-point grading system in which flavour and colour each constitute 40% of the grade, while 20% is for absence of defects; this is for products covered by US Standards for Grades of Orange Juice which include canned, frozen concentrated, reduced-acid frozen concentrated, canned concentrated, dehydrated and pasteurized orange juices, concentrated orange juices for manufacturing and orange juice from concentrate. Thus, even two samples of the same grade of orange juice may have different flavours!)*

(3) What possible flavouring compounds are in the flavour mix? *[Orange oil contains at least 112 volatile compounds. Of these, D-limonene is the most abundant (ca. 90–95%), but it is the oxygenated aldehydes and esters that are the*

most important contributors to orange flavour. Many oxygenated alcohols are present, but not ketones. Coumarins and flavonoids compose most of the non-volatile constituents (1–2%).[14] The major flavour compounds are listed for the following spices: vanilla (vanillin), ginger (gingerols, shoagols, cineol, borneol, geraniol), cloves (eugenol), cassia (cinnamaldehyde), nutmeg or mace (eugenol, cineol, terpineol, safrole, myristicin).]

(4) Are all 'natural extracts' really natural extracts? [A debate was started by a question from the instructor about the formation of vanillin from lignin (since lignin is from a plant source, then vanillin formed from the degradation of wood pulp should be a natural extract.) This led to a discussion of flavour compounds derived from turpentine. Are these natural? This led to close scrutiny of the US Code of Federal Regulations 21.CFR.101.22.a.3, and the category 'nature identical' used in the European Economic Community.]

Problem

You contact two other natural flavour suppliers, Bioglobex and Bioorbex, and send them a sample of the flavour mix that you currently use. You tell them that if they can match the flavour mix for both flavour and specifications from natural materials only, your company would buy some of their supplies from them.

After a few weeks, both of the flavour suppliers say that they cannot match the flavour mix by blending the natural extracts available to them: they could nearly achieve the flavour profile, but could not reproduce the combination of GC profile, percentage volatile oil, optical rotation and refractive index. Bioglobex want to know if you could change the specifications and Bioorbex want to use synthetic materials to be able to match the specifications.

Learning issues

(1) Why are these specifications required on the flavour mix? *(A manufacturer sets up specifications so that the incoming raw materials for his process are as consistent as possible.)*

(2) What is the procedure used to match the proprietary flavour mix? *(Flavour matching needs both the flavour chemist as well as the sensory analyst. Although the chemist identifies the individual chemicals, what is important is the flavour in the finished product. See Chapter 4.)*

Problem

You talk to your purchaser of ingredients and asked why the specifications are set as they are for the flavouring mix. He gives you an historical perspective: the current supplier had interacted with your predecessor when the product was first developed, and the specifications were probably written with guidance from the supplier. Thus, he really does not know why they are set as they are, but he uses the rule, 'If it ain't broke, don't fix it!' He has had no problems having his orders met quickly and accurately by this supplier.

You also talk with Biospherex. Why can they only supply 225 kg (500 lb) per year? Is it a limitation of their equipment? No, it seems that there was a huge hurricane that hit the major cassia producers, wreaked havoc with the current crop and destroyed some of the fields; these would take years to replant to production stage. Thus, the limit to production is the amount of the correct variety of cassia used in their blend. Cassia produced elsewhere gives a different flavour and GC profile than what is needed.

It has taken you over a month to realize that the problem is one of natural disaster! You decide that you need to go back to your boss and the marketing people and ask why it is so important to have all natural ingredients. You feel that only the addition of synthetic ingredients can solve the problem now. You are told that your company will lose a lot of customers because the label will have to contain either the chemical name or, at the very least, 'artificial flavourings'.

Well, you tell them that either the specifications and thus the flavour have to change slightly or the label has to change. The marketing people decide that a slight flavour change is better than a label change.

So you go back to Bioglobex and tell them to send the natural product that does not meet specifications so that you can test it. You make up some beverage and only 50% of the sensory panellists can tell a difference in a triangle test.

Prior knowledge

(1) Different varieties of spices give different flavour and GC profiles.

Learning issues

(1) What would happen if the specifications were changed or widened? *(Perhaps nothing if the specifications were too narrow to begin with. The question is how much inconsistency can be accepted in the raw material and a consistent finished product still be obtained? This should be tested with real samples and determined statistically. However, in industry, many times it is an intelligent guess by the developer of the product.)*

(2) What are the labelling laws (synthetic *vs.* natural)? *(The definition in 21.CFR.101.22.a.3 includes enzymatic and fermentation processes. Thus, biocatalytic but not chemical transformations of natural substances can be legally labelled as natural in the US.)*

(3) How is a triangle test done? *(In a triangle test, two samples are the same and one is different; the panellist must choose which of the three is the different sample.)*

(4) What is a significant difference in a triangle test? *(33% is chance. At the 90% confidence level, at least six out of nine people must choose the odd sample for it to be statistically significant. However, this really depends on the number of people and type of error you are willing to assume.)*

Problem

The sensory results are better than you expect, so you have the plant make up a batch with the flavour mix from Bioglobex and test market it in 10 stores locally. Only one complaint was registered at one store!

So, you rewrite the specifications to include either of the products. You decide to drop the refractive index and optical rotation specifications as they are the most inconsistent, and keep the percent volatile oil and GC fingerprint specifications, but just increase the allowable variation.

Learning issues

(1) Do the specifications control flavour variations? *[The percent volatile oil and GC profile are much closer indications of the flavour concentration and profile: percent volatile oil correlates to concentration of flavour and individual GC peaks can be correlated to type of aroma. The overall GC profile may be misleading when the major peaks have little flavour, as is the case for limonene in orange oil. Optical rotation and refractive index are quick and simple tests to check for consistency from one batch of a product to another, but are more indicative of the flavour carrier (alcohol, propylene glycol, lipid, emulsifier) than of the actual flavour compounds, because the bulk of the flavour mixture is actually non-flavourful compounds.]*

(2) Should the specifications have been changed? *(This is a matter of judgement, but generates good discussion. The consensus usually is that one should not be too hasty in a decision to rewrite the specifications. Customer complaints are not a good indication, sales figures over the long run are better; many consumers do not complain, they just vote with their pockets. Refractive index and optical rotation are historical tests used before chromatography became available and were primarily used to make sure a mistake was not made in the addition of ingredients. Since they are so much quicker than GC analysis, and can detect human errors, they should not have been so readily dropped; the specification could have simply been widened to include the second supplier's raw material. Perhaps these tests should be utilized as in-line monitors during processing, so human errors can be detected before a large amount of product is ruined. Adding a second supplier is an important step. In case of natural disasters, plant fires, or personnel strikes, it is always recommended to have more than one supplier of raw material.)*

MAILLARD REACTION

Problem

Your boss came back from Sweden where he had been served a delicious ethnic dish, which he describes to you as a meat-filled dumpling. He thinks a product of this kind would make a fantastic addition to your frozen dinners line: a meat-filled dumpling that could simply be reheated in a savoury broth. You talk to an elderly Swede in your neighbourhood and she tells you that they are delicious but usually take two days to make. Here is her recipe:

A beef or pork bone plus a few pounds of meat are cut into small pieces, blanched and simmered all day with carrots, celery, onions and potatoes to make the broth. For the evening meal, a little of the meat is served with the carrots, celery, onions and potatoes. The bone is discarded and the rest of the meat and the broth are cooled separately overnight. The next day, the fat that has solidified on top of the broth is removed. Then a yeast dough is made that is similar to bread, but is not as thick. While the dough is rising, the meat and a little of the broth are reheated along with some flour or cornflour. After the first rising, a spoonful of the meat and thickened sauce are placed inside a ball of dough and the dough is allowed to rise again. Just before serving, the meat-filled dough is dropped into boiling broth and boiled covered until the dumpling is not doughy.

Prior knowledge

(1) Soup is made the first day. The soup is flavoured by the carrots, celery, onions, potatoes, meat and bones.

(2) The Maillard reaction is the reaction between the proteins and reducing sugars that forms both the brown colour and the flavours during the browning of meats or the crust formation of baked goods.

(3) Meat is blanched to coagulate and remove blood proteins, so that the soup made subsequently is not cloudy. Flavours themselves are not lost here, but both the proteins and reducing sugars that could participate in the Maillard reaction are lost.

(4) Bones contain collagen that degrades into gelatin during the simmering of the soup, thus making the soup broth more nutritious. Collagen and gelatin proteins are flavourless, but could react to form flavours via the Maillard reaction.

(5) Removal of the solidified fat on the top of the broth results in the loss of fat-soluble flavour compounds.

(6) Yeast in the dumplings leads to fermentation and the production of alcohols.

Learning issues

(1) Are there any flavours that are not from the Maillard reaction in this process? *[(1) Yeast fermentation produces mainly alcohols which do not contribute much to flavour; however, with available free fatty acids they form small amounts of esters that are very flavourful. (2) There are vegetable flavours, such as β-ionone from the breakdown of β-carotene in the carrots, of phthalides and terpenoids in the celery, and of sulfur compounds from the onions.]*

(2) Where do Maillard reactions occur in the process of making meat-filled dumplings? *[(1) Soup simmered for a long time. (2) Dumplings boiled for a short time. (3) Sauce thickened over heat. The Maillard reaction can occur slowly in the presence of water, but different flavours are produced than if the food was at a higher temperature with less water. In the sauce, most of the water is tied up in the added starch thickener, so with less water there will be more flavour development. As water evaporates the Maillard reaction increases with time.]*

Problem

You decide to make a beef-filled dumpling. However, you know that the two day process cannot be used in your plant. Your company is already buying savoury powders to flavour sauces and is pressure-cooking cubed beef for other beef-based dinners. Thus, you feel that your challenge is to develop a dough that can be wrapped around the meat filling. Since your company does not make yeast doughs, and you also feel that making a yeast dough would take too long, you decide to look into quicker methods of making dumplings. You find a recipe in a cookbook for making dumplings using baking soda, flour, and buttermilk. You follow the recipe at home, but substitute whole milk which you had in the refrigerator and make them without any meat filling. Then you boil them in a bullion broth. However, the finished dumplings taste bitter!

Prior knowledge

(1) Buttermilk contains diacetyl which will participate in the Strecker degradation.
(2) The elimination of yeast reduces the amount of flavour in the dumplings.

Learning issues

(1) What caused the bitter flavour of the dumplings? *[There is a lack of acid in the whole milk; buttermilk has more acid. It is the Na_2CO_3 formed from reaction*

(5.1) that causes the bitter flavour.]

$$2NaHCO_3 \rightarrow CO_2 + Na_2CO_3 + H_2O \qquad (5.1)$$

(2) What are savoury powders? How are they made? *[Commercial savoury powders are Maillard reaction products, made by controlling the reaction of proteins and reducing sugars found in meat and seafood by-products. They are made by cooking the meat or seafood by-products and other additives in minimal water to extract the soluble proteins and sugars and allow the reaction to take place. The water is removed (spray-dried) before or after non-flavoured ingredients, such as salt or dextrose, are added to produce a free flowing powder.]*

(3) What is the difference between boiled beef and pressure-cooked beef? How does this affect flavour? *(Cooking under pressure is at a higher temperature than the boiling point of water. Thus, the Maillard reaction should proceed more quickly and some reactions with a higher activation energy will be able to proceed; but it may not proceed as completely. Time is important because the Maillard reaction continues on substrates present and those substrates are continually changing.)*

Problem

You decide that the bitterness was due to sodium carbonate formed because the whole milk did not contain enough acid. Thus, you repeat the dumpling recipe using baking powder instead of baking soda and produced something the whole family liked.

The next day at work, you follow the same recipe but use powdered milk from the plant (used in another product). The dumpling is browner and tastes strange! What happened?

You believe that the difference must be in the powdered milk. You obtain the specification sheet for the powdered milk and find that it is actually 'sweet dairy whey powder', a by-product in the manufacture of mozzarella cheese. Listed in Table 5.1 are the analyses shown on the specification sheet.

Prior knowledge

(1) Whey powder has a high concentration of lactose, which is a reducing sugar.

(2) The wheat proteins, gliaden and glutenin, form gluten with the mixing of the dough. These proteins can be utilized in the Maillard reaction.

Learning issues

(1) How do leavening ingredients affect flavour? *(Baking soda and powder*

Table 5.1 *Minimum ingredient specification for sweet dairy whey*

Protein	11% min.
Moisture	5% min.
Lactose	75% min.
Fat	1.5% min.
Ash	9.9% min.
pH	5.6 min.
Titratable acidity	0.16% min.

raise pH which affects flavour-development reactions. The optimal range for the Maillard reaction is between pH 4.5 and 6.5. If the dough is too alkaline, $NaHCO_3$ forms Na_2CO_3, which is bitter. Yeast forms glucose diphosphate which breaks down quickly during cooking into a dicarbonyl that can participate in the Strecker degradation.)

(2) What are the differences between powdered non-fat dry milk (NFDM) and whey powder? How does this affect flavour? *(Powdered milk is whole milk with the fat removed; whey powder is a by-product of cheese making. Whey proteins are mainly β-lactoglobulin and α-lactoalbumin and are very low in caseins. When the type of protein changes, it will react differently in the Maillard reaction, thus producing different flavours.)*

Problem

Since the price of whey powder is one-third the cost of NFDM, you talk to your supplier and find that for 1/2 the cost of NFDM, he can supply you with lactose-reduced whey powder. This product arrives just as pressure-cooked beef is being made in the plant, so you decide it is time to do a controlled experiment. You make the dumplings four different ways:

A. Baking powder with NFDM.
B. Baking powder with lactose-reduced whey powder.
C. Yeast with NFDM.
D. Yeast with lactose-reduced whey powder.

You consider C to be your control as it is essentially the original recipe for the dough for the Swedish meat-filled dumpling. For both C and D, you wrap the dough around the beef and allow the dough to rise for the final proofing. Then you cook it in boiling broth. For A and B, you simply wrap the dough around the beef and cook it immediately in boiling broth.

Your laboratory staff tasted all the dumplings and found the flavour of all of them quite good, but felt that both yeast dumplings had a better flavour and were much more visually appealing, as the shape was uniform and the meat was more centred in the dough. The slight difference in flavour between the two milk powders was not considered significant.

Finally, you served your boss the dumplings. When he says that C was just what he was thinking of, you ask him if he is ready to go into the bakery business as yeast dough technology is currently foreign to his plant!

Prior knowledge

(1) From the problem, a difference in proteins, caseins versus lacto-globulins and lactoalbumins, produced only a slight difference in the flavour. Thus, a combination of the two milk powders should be used for the next experiments until no difference can be detected between the NFDM and mixture by a triangle test. This will cut costs without affecting flavour.

Learning issues

(1) How is the lactose removed from the whey? *(Membrane concentration of the whey proteins removes small molecules, including lactose, with the water.)*

THIO-STENCH

Problem

Your friend, Susan, was cooking sauerkraut on her gas stove indoors. At the same time her husband, Steve, was using the propane grill outside to fry sausages. This was the first warm spell after a long cold winter, and they had invited you over for dinner. When you walk in the door, you almost turn back as you smell a horrible stench. You yell at Susan and ask what was stinking up her home and she hollers back from the kitchen 'What smell? Oh, you probably mean the sauerkraut!' 'No,' you shout, 'it doesn't smell like sauerkraut'. You walk through the house to the kitchen and tell Susan that the smell seems to be coming from her basement. You call to Steve, but he is talking to his neighbour who is a mushroom grower. They both come inside and insist that you are joking with them. By this time, you are wondering; you can hardly smell it yourself!

Prior knowledge

(1) Adaptation is occurring.
(2) Thio means sulfur compounds, which are quite pungent and have a very low olfactory threshold.

(3) Microbial growth (fermentation) and oxidation can cause strong aromas.

(4) Temperature changes can cause problems.

(5) Mercaptans (thiols) are added to both natural gas and propane so that you can smell any leaking gas.

Learning issues

(1) What is the mechanism of adaptation? *(There are three: a decrease in the number of receptor molecules, the depletion of receptor transmitter substances, and feedback from the central nervous system that dampens the receptor processes.)*

(2) What are the possible sources of sulfur compounds in this problem?
   a. Sauerkraut. *(Glucosinolates transform to isothiocyanates in raw cabbage and then to some 1-cyano-4-methylsulfinylbutane and 1-cyano-3-methylsulfinyl-propane via fermentation.)*
   b. Sausage. *(H_2S, CH_3SH, CH_3SCH_3, CH_3SSCH_3, thiophenes, thiazoles, thiazolines.)*
   c. Mushroom soil. *(H_2S, CH_3SH, CH_3SCH_3, CH_3SSCH_3.)*
   d. Leaking natural gas or propane. *(Mercaptans added.)*
   e. Microbes, moulds and fungus grow better in basements when warm. *(Pseudomonas produces CH_3SH, CH_3SCH_3, CH_3SSCH_3.)*
   f. Sewer in basement *(Sewer gas contains H_2S and CH_4.)*

(3) What types of reactions occur when sulfur-containing compounds found in foods are grilled (sausage) and boiled (sauerkraut)? *(Heat produces less alkyl sulfides and more cyclic sulfur compounds.)*

(4) How do temperature and humidity affect the basement? *(Microbes grow when humidity is high; a sewer trap also dries out when humidity is low during the winter.)*

(5) Are there interactions among the aromas from the different sources? *(Certainly; however, adaptation is probably more important.)*

(6) How do weather conditions influence the perception of odours? *(Humidity keeps the membranes in the nose moist and affects the ability of volatile molecules to reach the olfactory neurons, and temperature affects the volatility of flavour compounds.)*

This problem developed from a conversation with an engineer at a power company. There seem to be a lot of 'false alarms' of gas leaks that the company has to investigate due to food aromas, transferring of mushroom soils, and mouldy basements. The one that hits the first warm spell every spring is due to sewer gas. The traps that keep the aromas in the pipes are normally filled with water. During the winter, these pipes dry out because of low humidity and non-use. When the temperature rises this enables microbial fermentation in the pipes, and the gas company is flooded with phone calls!

REFERENCES

1. R.L. Rouseff, ed., *Bitterness in Foods and Beverages*, Developments in Food Science Series, Vol. 25, Elsevier, Amsterdam, 1990.
2. G.A. Burdock, ed., *Fenaroli's Handbook of Flavor Ingredients*, CRC Press, Boca Raton, 3rd edn, 1995.
3. K. Kurihara, Y. Kurihara and L.M. Beidler. In *Olfaction and Taste*, ed. C. Pfaffmann, Rockefeller University Press, New York, 1969, 450.
4. R.F. Margolskee. In *Handbook of Olfaction and Gustation*, ed. R.L. Doty, Marcel Dekker, New York, 1995, p.575.
5. D.R. Burgard and J.T. Kuznicki, *Chemometrics: Chemical and Sensory Data*, CRC Press, Boca Raton, 1990.
6. J.E. Amoore, *Molecular Basis of Odour*, The Bannerstone Division of American Lectures in Living Chemistry, No. 773, ed. I.N. Kugelmass, Charles C. Thomas, Springfield, 1970.
7. A.N. Gilbert and C.J. Wysocki, *National Geographic*, 1987, **172**, 514.
8. S. Takafuji. In *Food Flavours, Ingredients and Composition*, Vol. 32, ed. G. Charalambous, Elsevier, Amsterdam, 1993, 175.
9. M. Boelens, P. de Valois, H. Wobben and A. van der Gen, *J. Agric. Food Chem.*, 1971, **19**, 984.
10. H. Maarse, ed., *Volatile Compounds in Foods and Beverages*, Marcel Dekker, New York, 1991.
11. M. Kamm. In *Enclycopedia of Food Science and Technology*, Vol. 3, ed. Y.H. Hui, Wiley, New York, 1992, **3**, 1946.
12. O.R. Fennema, *Food Chemistry*, 2nd edn., Marcel Dekker, New York, 1985.
13. P.A. Tice, *Food Additives & Contaminants*, 1994, Mar/Apr, 187.
14. G. Dugo, *J. Essent. Oil Res.*, 1994, **6**, 101.

Bibliography

GENERAL

Bessiere, Y. and Thomas, A.F., eds., *Flavour Science and Technology*, Wiley, New York, 1990.

Birch, G.G. and Lindley, M.G., eds., *Developments in Food Flavours*, Elsevier Applied Science, New York, 1986.

Boudreau, J.C., ed., *Food Taste Chemistry* ACS Symposium Series, American Chemical Society, Washington, DC, 1979, No. 115.

Burdock, G.A., ed., *Fenaroli's Handbook of Flavor Ingredients*, CRC Press, Boca Raton, 3rd edn., 1995.

Charalambous, G., ed., *Food Flavors, Ingredients and Composition*, Developments in Food Science Series, Elsevier, Amsterdam, 1993, Vol. 32.

Charalambous, G., ed., *Off-Flavors in Foods and Beverages*, Developments in Food Science Series, Elsevier, Amsterdam, 1992, Vol. 28.

Heath, H.B. and Reineccius, G., *Flavor Chemistry and Technology*, AVI Publishing Company, Westport, 1986.

Hui, Y.H., ed., *Encyclopedia of Food Science and Technology*, Wiley, New York, 1992.

Maarse, H., ed., *Volatile Compounds in Foods and Beverages*, Marcel Dekker, New York, 1991.

Morton, I.D. and Macleod, A.J., eds., *Food Flavours; Part A. Introduction*, Developments in Food Science Series, Elsevier, Amsterdam, 1982, Vol. 3A.

Morton, I.D. and Macleod, A.J., eds., *Food Flavours; Part C. The Flavour of Fruits*, Developments in Food Science Series, Elsevier, Amsterdam, 1990, Vol. 3C.

Reineccius, G., ed., *Source Book of Flavors*, 2nd edn; Revised book originally by Henry Heath ed., Chapman & Hall, New York, 1994.

Teranishi, R., Flath, R.A., and Sugisawa, H., eds., *Flavor Research: Recent Advances*, Marcel Dekker, New York, 1981.

BIOLOGY

Doty, R.L., ed., *Handbook of Olfaction and Gustation*, Marcel Dekker, New York, 1995.

Getchell, T.V., Doty, R.L., Bartoshuk, L.M. and Snow, J.B., eds., *Taste and Smell in Health and Disease*, Raven, New York, 1991.

Serby, M.J. and Chobor, K.L., eds., *Science of Olfaction*, Springer-Verlag, New York, 1992.

BIOTECHNOLOGY

Goldberg, I. and Williams, R.A., eds., *Biotechnology and Food Ingredients*, Van Nostrand Reinhold, New York, 1991.

Mor, J.-R., in *Flavour Science and Technology*, eds. Y. Bessiere and A.F. Thomas, Wiley, New York, 1990.

Patterson, R.L.S., Charlwood, B.V., MacLeod G. and Williams, A.A., eds., *Bioformation of Flavours*, The Royal Society of Chemistry, Cambridge, 1992.

Parliament, T.H., and Croteau, R. eds., *Biogeneration of Aromas*, ACS Symposium Series, American Chemical Society, Washington DC, 1986, No. 317.

Schrier, P., ed., *Bioflavour '87: Analysis, Biochemistry, and Biotechnology*, Walter de Gruyter, Berlin, 1988.

FLAVOUR CHEMISTRY

Acree, T.E. and Soderlund, D.M. eds., *Semiochemistry Flavors and Pheromones*, Proceedings ACS Symposium, Washington DC, Walter de Gruyter, Berlin, 1983.

Ellis, J.W., Overview of Sweeteners, *J. Chem. Ed.*, 1995. **72**, 671.

Kjaer, A. In *The Biology and Chemistry of the Cruciferae*, eds. J.G. Vaughan, A.J. MacLeod and B.M.G. Jones, Academic Press, London, 1976, 207.

Fenwick, G.R., Heaney, R.K. and Mullin, W.J. 'Glucosinolates and their breakdown products in food and food plants' *CRC Critical Reviews in Food Science and Nutrition*, 1983, **18**, 123.

Maga, J. In *Phenolic Compounds in Food and their Effects on Health I: Analysis, Occurrence, and Chemistry*, eds., Ho, C.-T., Lee, C.Y. and Huang, M.-T., ACS Symposium Series, No. 506, American Chemical Society, Washington DC, 1992.

Ohloff, G. In *Progress in the Chemistry of Organic Natural Products*, Vol. 35, eds. Herz, W., Grisebach, H. and Kirby, G.W., Springer-Verlag, New York, 1978, p. 431.

Teranishi, R., Takeoka, G.R. and Guntert, M. eds., *Flavor Precursors: Thermal and Enzymatic Conversions*, ACS Symposium Series, Vol. 490, American Chemical Society, Washington, DC, 1991, Vol. 490.

Lindsay, R.C. In *Food Chemistry*, ed. Fennema, O.R., Marcel Dekker, New York, 1985, p. 585.

Rouseff, R.L. ed., *Bitterness in Foods and Beverages*, Developments in Food Science Series, Vol. 25, Elsevier, Amsterdam, 1990, **25**.

INSTRUMENTAL ANALYSIS

Ho, C.-T. and Manley, C.H., eds., *Flavor Measurement*, IFT Symposium Series, Marcel Dekker, New York, 1993.

Maarse, H. and Belz, R., *Isolation, Separation and Identification of Volatile Compounds in Aroma Research*, D. Reidel, Dordrecht, 1985.

Middleditch, B.S., *Analytical Artifacts*, Elsevier, Amsterdam, 1986.

Nielsen, S.S., *Introduction to the Chemical Analysis of Foods*, Jones and Bartlett, Boston, 1994.

Vandeweghe, P. and Reineccius, G., *J. Agric. Food Chem.*, 1990, Vol. 38, 1549.

PROBLEM-BASED LEARNING

Albanese, M. and Mitchell, S., *Acad. Med.*, 1993, Vol. 68, 52.

Boud D., and Feletti, G., *The Challenge of Problem Based Learning*, St. Martin's Press, New York, 1991.

SENSORY

Burgard, D.R. and Kuznicki, J.T., *Chemometrics: Chemical and Sensory Data*, CRC Press, Boca Raton, 1990.

Civille, G. and Lyon, B., *Aroma and Flavor Lexicon for Sensory Evaluation*, ASTM, West Conshohocken, 1996.

Hootman, R.C., *Manual on Descriptive Analysis Testing for Sensory Evaluation* ASTM, West Conshohocken, 1992.

Lawless, H.T. and Klein, B.P., eds., *Sensory Science Theory and Applications in Foods*, IFT Basic Symposium Series, Marcel Dekker, New York, 1991.

McBride, R.L. and MacFie, H.J.H., *Psychological Basis of Sensory Evaluation*, Elsevier Applied Science, New York, 1990.

Meilgaard, M., Civille, G.V. and Carr, B.T., *Sensory Evaluation Techniques*, CRC Press, Boca Raton, 1987, Vol. I and II.

Piggott, J.R., *Sensory Analysis of Foods*, Elsevier Applied Science, New York, 2nd edn, 1988.

Stone, H. and Sidel, J., *Sensory Evaluation Practices*, Academic Press, New York, 1985.

Vries, H.D., and Stuiver, M. In *Sensory Communication*, ed. W. Rosenblith, M.I.T. Press, Cambridge, 1961, 159.

Yantis, J.E., *The Role of Sensory Analysis in Quality Control*, ASTM, West Conshohocken, 1992.

Glossary

accessory olfactory bulb – a secondary bulb, medial to the main olfactory bulb, that specializes in processing pheromones.

adaptation – a reversible decrease of the sensitivity of an individual to a compound immediately after contact with the compound.

adenylate cyclase (AC) – an intermediary enzyme in the transduction process that permits the reduction of adenosine triphosphate to cyclic adenosine monophosphate.

agglomeration – a mass of particles clustered together.

aglycone – the non-sugar part of a glycoside.

aliphatic – organic chemical term for non-aromatic organic molecules.

amygdala – a target of the lateral olfactory tract, from which axons continue to the medial hypothalamus, perhaps to mediate odour-induced fear.

anterior insula – part of the primary taste region of the cortex.

anterior olfactory nucleus – a target of the olfactory tract and the main source of connections to the opposite olfactory bulb.

astringency – the property of a substance to contract the tissues or canals of the body. Astringent flavours provide a sensation of puckering inside the mouth.

axon hillock – the point where the axon leaves the cell body of a neuron.

azeotrope – a mixture of two or more compounds that has a fixed boiling point without decomposition and boils off at a constant ratio of the compounds.

basal cells – small cells at the base of the taste bud or olfactory epithelium from which new receptors develop.

body wisdom – the notion that animals consume that which is good for them and avoid chemicals that are toxic.

carnosine – the dipeptide, N-β-alanyl-L-histidine, which is a neuromodulator that influences the action of glutamate in the olfactory bulb.

caudolateral orbitofrontal cortex (OFC) – secondary taste cortex, dealing with highly-processed information.

cephalic phase reflexes – digestive reflexes that have their origins in the brain.

chemometrics – the use of multiple variables in mathematics to determine correlations between chemical data and physical phenomena.

chloroplasts – green vesicles in plant cells where photosynthesis takes place.

chorda tympani nerve – the main taste nerve from the front two-thirds of the tongue.

chromoplasts – storage vesicle for carotenoids in a plant cell.

circumvallate papillae – 7–11 prominent swellings towards the back of the tongue in whose trenches are contained several hundred taste buds.

classification – systematic grouping of a lot based on similar characteristics or traits such as particle size, gravity, density, *etc.* (Generally used in regards to particulate solids.)

climacteric fruits – those fruits that go through a critical period when added ethylene will act as a hormone to hasten ripening.

conditioned taste aversion – the revulsion that develops to a taste that has been associated with nausea.

cordial – a strong, sweetened, aromatic alcoholic liquor.

cribriform plate – the fenestrated portion of the ethmoid bone through which the olfactory nerve passes from the epithelium to the bulb.

cyclic adenosine 3′, 5′-monophosphate (cAMP) – the second messenger that opens sodium channels within a receptor cell to cause activation.

dark cells – taste receptor neurons, thought to be early in their life cycle.

detection threshold – the lowest concentration at which humans can perceive the presence of a compound.

dopamine – 4-(2-aminoethyl)benzene-1,2-diol, a neurochemical whose release is associated with feelings of pleasure.

dorsal motor nucleus of the vagus (DMNX) – the nucleus whose neurons send projections to the gut to control autonomic functions.

empyreumatic – being transformed in fire.

enantiomer – a compound that is optically active whose mirror image rotates plane polarized light in the opposite direction.

external plexiform layer – the third layer of the olfactory bulb. It is composed primarily of the dendrites of mitral, tufted and granule cells.

facial nerve – the seventh cranial nerve, which houses the chorda tympani nerve.

flavour antagonism – when one or more compounds are mixed together and interfere with flavour detection, *i.e.*, the response (actual) is less than the arithmetic sum (theory).

flavour synergism – when one or more compounds are mixed together and enhance the ability to detect the flavour, *i.e.*, the response (actual) is greater than the arithmetic sum (theory).

foliate papillae – five or six folds on the lateral surface of the epithelium of the tongue in each of which are contained more than one hundred taste buds.

free radicals – molecules which contain unpaired electrons and thus are very reactive and short-lived.

frontal operculum – part of the primary cortical taste area.

functional magnetic resonance imaging (fMRI) – a technique similar to NMR for determining which brain areas are active as the subject experiences a specific stimulus.

fungiform papillae – mushroom-shaped mounds that cover the anterior two-thirds of the tongue. They contain taste cells that are particularly sensitive to salty and sweet stimuli.

geniculate ganglion – the collection of cell bodies whose axons form the chorda tympani taste nerve.

glomerular layer – the second layer of the olfactory bulb. This is the locus of interaction between first- and second-order olfactory cells.

glossopharyngeal nerve – the ninth cranial nerve, carrying taste information from the foliate and circumvallate papillae, and the pharynx and palatoglossal arches.

glutamate – the most common excitatory neurotransmitter in the olfactory bulb.

glycoprotein receptive sites – the receptor molecules, made up of proteins with attached sugars, to which specific odorants bind.

glycoside – a molecule made up of a non-sugar part and a sugar.

granule cells – small, numerous cells whose dendrites mediate local interactions among mitral cells.

granule cell layer – the sixth (innermost) layer of the olfactory bulb. Its granule cells modulate the activity of mitral cells.

greater superficial petrosal nerve – the nerve that serves taste receptors on the soft palate.

guanosine triphosphate (GTP)-binding protein – molecules released when an odorant binds to an olfactory receptive site. The release of GTP-binding proteins within the receptor cell amplifies the olfactory signal.

hedonic – having to do with pleasure, preference judgements (like–dislike).

hedonic scale – line scale used in sensory testing (5 pt = very bad, bad, neutral, good, excellent) which is not linear.

hypothalamus – region of the brain concerned with emotions and motivation, including the relationship of the chemical senses to feeding and hedonic processes.

intermediate cells – taste receptor cells, thought to be in the midst of their twelve-day life cycle.

internal plexiform layer – the fifth layer of the olfactory bulb. It is composed mainly of axons and axon collaterals from mitral and tufted cells.

kinaesthesis – the sensation of bodily position, presence, or movement resulting chiefly from stimulation of sensory nerve endings in muscles, tendons and joints.

labelled-lines – specific coding channels assigned to signal the presence of each basic taste quality.

lachrymator – a substance that causes the shedding of tears.

lachrymatory – the property of a substance to cause the shedding of tears; *e.g.* tear gas.

lateral inhibition – the process by which distinctions among olfactory signals are sharpened as activity in one glomerulus suppresses the responses in those that surround it. Lateral inhibition is mediated by periglomerular cells in the olfactory bulb.

lateral olfactory tract – the major pathway from the olfactory bulb to the olfactory cortex. It is composed of the axons of mitral and tufted cells.

light cells – taste receptor cells thought to be towards the end of their life cycle.

limbic system – a collection of brain areas involved in the control of emotion.

macerate – to soften or decompose a food by the action of a solvent.

macrosmatic – having a large and well-developed olfactory system.

microsmatic – having a small or poorly developed olfactory system.

microvilli – the hair-like extensions of receptor cells on which are located the binding sites for odorant molecules.

mitral cells – the large, prominent neurons that receive information from the olfactory nerve and project their axons to the olfactory cortex.

mitral cell layer – the fourth layer of the olfactory bulb. It is composed of the cell bodies of mitral cells.

neuron types – the division of taste and olfactory cells into discrete classes, each associated with a particular basic quality.

nodose ganglion – collection of cell bodies whose axons form the vagus nerve.

non-climacteric fruits – those fruits whose ripening cycle is unaffected by ethylene.

non-volatile – does not evaporate readily.

nucleus of the solitary tract (NST) – the first central relay for taste information and for input from the gut.

odorant-binding proteins (OBPs) – proteins that bind to specific odour molecules and transport them to the receptor surface.

olfactory bulb – the main information processing station of the olfactory system. This six-layered structure receives olfactory information from the epithelium, and sends highly-processed information to the olfactory cortex.

olfactory cortex – any of the several cortical regions that receive projections from the olfactory bulb. These include the anterior olfactory nucleus, the piriform cortex, the olfactory tubercle, the amygdala, and the temporal entorhinal cortex.

olfactory epithelium – the receptor surface area where odorant molecules are bound. Its neurons form the olfactory nerve, which projects to the olfactory bulb.

olfactory nerve layer – the first (most superficial) of the six layers in the olfactory bulb. It is composed of incoming fibres from the olfactory epithelium.

olfactory placode – in embryonic development, the pit in which the olfactory epithelium develops.

papillae – small mounds on the tongue that house anywhere from zero to several hundred taste buds. They are of four types: filiform, fungiform, foliate and circumvallate.

parabrachial nuclei (PBN) – a relay in the pons for autonomic information from the viscera (internal organs).

pattern theory – theory of sensory coding that asserts that each sensory cell is involved in every signal according to its contribution to an overall pattern of activity.

periglomerular cells – small neurons that lie at the base of olfactory glomeruli. They are activated by input from the olfactory nerve, and send a wave of inhibition to adjacent glomeruli to accentuate differences in olfactory signals.

petrosal ganglion – collection of cell bodies whose axons form the glossopharyngeal nerve.

phase separation – the segregation of two immiscible phases with time (two liquids, liquid–gas, liquid–solid); antonym: emulsification.

pheromones – molecules that carry olfactory information related to a particular species. Much of this information has to do with reproduction.

piriform cortex – the main projection area of the lateral olfactory tract. Its pyramidal cells project olfactory input to the mediodorsal thalamus for identification.

pungency – the property of a substance to give an acutely distressing feeling. Pungent flavours are biting, sharp, and acrid.

pyramidal cells – the most prominent cells in piriform cortex. They receive axons from the lateral olfactory tract and send projections to the mediodoursal thalamus.

pyrexia – the abnormal elevation of body temperature.

recognition threshold – the lowest concentration at which we can recognize the flavour of the compound we detect.

reverted oil – processed and deodourized vegetable oil that has been stored incorrectly or for too long, in which the fatty acids have become oxidized and the oil has *reverted* to having an odour again.

sodium appetite – a strong desire for sodium resulting from the depletion of the body's reserves of salt.

short axon cells – a collection of neurons whose dendrites project among the glomeruli of the olfactory bulb.

stratification – segregation into segments that varies with respect to the property under study. (Generally used in regards to fluid samples.)

supporting or sustentacular cells – rectangular cells located in the upper one-third of the olfactory epithelium that provide structural support.

tactile receptor – an end organ of the sensory neurons that is stimulated by touch; thus where the sense of touching an object originates.

taste buds – goblet-shaped structures each containing about 100 cells among which are taste receptor neurons.

taste pore – the small opening at the top of each taste bud, through which the microvilli of taste cells protrude to sample the chemical environment.

temporal entorhinal cortex – part of the olfactory cortex, it sends projections to the hippocampus, perhaps to mediate olfactory memories.

terpenoids – naturally occurring hydrocarbons (the terpenes) and their oxygenated derivatives, which are based structurally and biosynthetically on the condensation of isoprenoid or isopentyl C_5-units.

topographic organization – the relationship between a taste or olfactory quality and the physical location at which it is represented in the nervous system.

trigeminal – pertaining to the chief facial sensor nerves. Usually refers to the pain sensations felt as heat or bite from flavour compounds, but these nerves also transmit mouthfeel.

trigeminal nerve – the fifth cranial nerve, through which the chorda tympani nerve leaves the tongue.

tufted cells – cells located in the external plexiform layer of the olfactory bulb. They receive input from the olfactory epithelium, and send their axons to the olfactory cortex in parallel with those from the mitral cells.

turbinates – folds in the olfactory epithelium that offer great surface area for olfactory receptors.

umami – candidate to be a primary taste stimulus. Umami is also associated with flavour enhancement. Represented by monosodium L-glutamate (msg). There is a controversy over whether umami is a basic taste like sweet, salty, sour and bitter. Umami has been described as a savoury flavour that makes foods taste good. The word comes from the Japanese meaning 'deliciousness' or 'savouriness'.

vagus nerve – the tenth cranial nerve, carrying taste information from the extreme posterior tongue, the oesophagus and the epiglottis.

ventroposterior medial nucleus pars parvocellularis (VPMpc) – the relay nucleus for taste in the thalamus.

viscera – the internal organs.

visceral – having to do with the gut and autonomic nervous system.

volatile – evapourating rapidly; changing readily into the form of vapour.

vomeronasal organ – secondary olfactory receptor, specialized for the detection of pheromones.

von Ebner glands – salivary glands of the tongue whose secretions bathe the papillae.

Subject Index